"CAMOUFLAGE THROUGH LIMITED DISCLOSURE"*

Deconstructing a Cover-Up of the Extraterrestrial Presence

*(a quote from Philip J. Corso, author of *The Day After Roswell*)

by

Randy Koppang

THE BOOK TREE
San Diego, California

ISBN 978-1-58509-110-2

Printed on Acid-Free Paper

Published by
The Book Tree
P O Box 16476
San Diego, CA 92176
www.thebooktree.com

Table of Contents

Acknowledgment

In 1989, former NASA Mission Specialist Robert Oechsler released a project report: *The Chesapeake Connection, An Implication of Corporate Involvement in the Cover-up!* This document sets forth a standard of analysis uncommon in ufology, and it firmly establishes a role model of investigation, collectively performed by The Annapolis Research Group.

I hereby acknowledge an honorable mention of Mr. Oechsler and his collaborators. Truly groundbreaking insights were deduced, and their contribution to ufology was unprecedented for the time. More importantly, *The Chesapeake Connection* anticipated precisely their stated "implication," as disclosed many years later by "insider" sources—namely, those quoted here.

There is legitimacy in this recognition: the fact that ufology per sé has yet to prioritize how *The Chesapeake Connection* had the potential to "forever change the course of ufological investigations." Then, nine years later, Col. Philip J. Corso and Bill Uhouse both verified Oechsler's pattern recognition with their firsthand witness accounts. But ufology still has yet to reconcile the learning-curve that ufology itself is instrumental in...

Thus, as Col. Corso said so clearly about the military's management of the ET-presence: "*...We never hid the truth from anybody, we just camouflaged it. It was always there, people just didn't know what to look for or recognize it for what it was when they found it. And they found it over and over again*" (page 78).

Yes, and so it shall be till it finally sinks in...

Introduction

Fifty-nine years of data collection has been applied to the enigma of extraterrestrials (ETs) visiting earth. A resulting pattern of evidence emerges from this collective inquiry. This data pattern indicates a host of complicit individuals who are shown to be possessing affirmative extraterrestrial (ET) evidence. Identifying these people has indicated the enclave of their careers, the military-industrial complex. Residing within this complex is a system for managing ET evidence: a contingency for popularizing ETs. The most persuasive account of this scheme is detailed in *The Day After Roswell* by Col. Philip J. Corso.

The cases presented in this book are internally consistent, and reinforce the assertions of Col. Corso. These independent voices of experience are Melinda Leslie and William "Bill" Uhouse. Together the combined body of data from these three sources gives cause to rethink the "cover-up" model of why ET evidence is undisclosed. Col. Corso has already explained the *management* of ET knowledge, in his chapter entitled *The Strategy.* It is my contribution here to begin reconciling Corso's unofficial "disclosure."

In rethinking a "cover-up" model for investigating ETs, the following questions arise:

1. Why would the "cover-up" actually disguise its intended purpose?

2. Has the UFO "cover-up" been simply a model of social acclimation, requiring a learning-curve of lengthy investigation to *perceive* it?

3. Is the ET-presence a human phenomena before it's an ET phenomena?

4. If an Official Disclosure of ETs was announced, would National Security restrictions limit our "need to know" to what we already do know?

5. Is the bottom-line of ET information an issue of economic dominance?

PART ONE

Overview

"...the censors were out on a longer limb than I knew. Our discovery of this was more than another good break—it was a coincidence so incredible that no fiction writer would dare to use it." –Major Donald E. Keyhoe, *Flying Saucers: Top Secret, 1960*

For the purpose of providing UFO studies with comprehensive perceptions, the testimonies of Melinda Leslie and Bill Uhouse are published here in tandem. The ufological insights reported by these two witnesses were made available independently. Their evidence has been combined because their data brings complimentary and integral comprehension to definitive historical circumstances; namely, the human response to UFO/ET phenomena in the form of Military Industrial Complex programs. Such human-scale relevance is something humans can easily appreciate. And exploring such human pursuits led Ms. Leslie to deduce an M.I.C. objective: The "reverse-engineering of" certain capabilities "the ET-abductee" experience provides them with.

The intent here is to specifically discern a data-basis for making correlations, and those of human activities generally. Especially, the circumstances affording political implication.

Of previously unavailable value to our associations here, is the disclosure included by Bill Uhouse. Dr. Steven Greer featured a brief contribution to his *Disclosure Project* document (2001) by Bill Uhouse. Our presentation, however, is the longest and most extensive condensation of the professional experience attributed to Uhouse: a career-long involvement with the engineered simulation of ET-disc craft.

Melinda Leslie's body of evidence is the product of her experience, plus continuing investigations. Leslie's focus is upon a subset of the ET-abduction phenomena. It involves the monitoring and interrogation of ET "experiencers," by *human* special operations agents. The Uhouse case achieves what no other known discloser, to date, has done so coherently: he synergizes with the internal logic in Leslie's evidence for human information retrieval *from* ET-abductees. With political implication, the converging contextual motives deduced in Melinda's evidence are consistent with M.I.C. pursuits reported by Uhouse. In other words, Uhouse does not link these two scenarios of data/experience intentionally. Yet, he explains an R&D process logically deduced from Leslie's evidence, overall. This is where an inter-connectedness between the two cases converge. The combined value of insight is, thus, much greater than the sum of these two cases, which exist independently.

The utilitarian settings for the alleged experience of both Uhouse and Leslie are the same: military-industrial projects. And, when linked through analytical correlation, Leslie and Uhouse validate the reverse-engineering strategy explained by Col. Philip J. Corso in his book *The Day After Roswell*. Thus, Uhouse, Leslie and Corso validate each other independently.

Finding a Model that Explains All the Data

With plenty of political resonance to go around in these cases an additional set of correlations has been recently reported (May 2005) by Linda Moulton Howe.

L. M. Howe's recent report magnifies the political sustenance of all the above by many orders of magnitude! Her report details a unique investigation into UFO crash retrieval cases. The "official research" was conducted by former Las Cruces, New Mexico State Representative Andrew Kissner.

While still in office, Rep. Kissner began confirming the United States' complicity in military/UFO hostilities. These clashes all occurred during the late 1940s and early 1950s period. Linda Howe reports: the conclusions drawn by Kissner were based on "off the record conversations with military and intelligence operatives... and government documents leaked to him." Rep. Kissner also confirms a Truman Administration directive to "collect a specimen" of the unidentified discs: the method was to intentionally shoot them down! In the late '40s this was done with the first land-to-air missiles; later, a high-intensity microwave radar was deployed. In triangulation, these radar beam-sites were built and used in New Mexico. This newest set of historical factors would clearly remove any doubt about the fully conscious motives, which instituted National Security policies for protecting assets retrieved from the flying disc phenomena.

Representative Kissner has written a yet-to-be published manuscript about his research. Meanwhile, on Kissner's behalf, Linda M. Howe reports the soundness and credibility of Kissner's correlations with current UFO studies in other areas—the obvious case being *MJ 12* document inferences; from crashed disc retrievals to reverse-engineering. Rep. Kissner is apparently documenting a major contribution toward apprehending the full magnitude of the UFO crash retrieval controversy: i.e., "several" saucers came down because humans were shooting them down! These possible facts would entirely change historical connotations of both UFO and conventional post-WWII history. Such an explanation for the enigma of saucers crashing would make sense. The military minds would certainly try to achieve a shootdown if possible.

In April 2005, Linda M. Howe publicly lectured about this alleged shootdown policy. Howe quoted Rep. Kissner on the first deliberate effort by the U.S. military to shoot down a "flying disc" at White Sands Proving Ground, New Mexico. Kissner determined that this occurred on May 15, 1947. On that day, the crash of a V-2 rocket test apparently resulted from a "radar target," observed as interfering with the V-2 flight path. "And immediately the V-2 rocket changed course." The "radar target" was reported as actually an "unidentified aerial disc," causing the V-2 to veer off course and abortively crash.

Howe continued Rep. Kissner's account by saying that "an Army officer at White Sands Proving Ground told Kissner the *peculiar phenomena* object was defined, then, as hostile—in that it appeared to have affected the V-2 rocket's trajectory...and high priority was assigned to collect a specimen of the technology for further analysis. And an emphasis was made on doing this covertly."

Research Comes Full Circle With History

Again, during a V-2 rocket launch on 29 May 1947, "flying discs" reappeared. The military successfully shot down one of them by first damaging it with a "proximity fuse" missile. The disc then flew south only to self-destruct near Juarez, Mexico. However, Rep. Kissner's account of these two V-2/disc incidents is actually what amounts to "the rest of the story"—because several V-2/disc sightings were credibly documented by ufologists in that time, where "aerial discs" were clearly observed *pacing* V-2 launches at White Sands, New Mexico. Another important case of such V-2/disc incidents will be cited below. It will be offered in a different but related setting according to the late astronaut, Gordon Cooper. Expecting no physical evidence, our purpose here is to establish whether all this is internally consistent. And it is! The question, however, is how or why might a history of Human/ET hostilities be used to disadvantage ufology?

Ironically, the most basic elements of this nearly official disc-shootdown history were already documented by ufologists contemporary with the events. Now, thanks to Rep. Kissner, the total body of data comes full circle. Credible specifics about an aggressive *shootdown* policy were all that were lacking. And this "specimen" motivated policy is necessarily indispensable for placing the reverse-engineering disclosure here, by Bill Uhouse, in logical perspective.

Linda Howe quotes researcher Leonard J. Stringfield on the shootdown policy: "We had personal knowledge about the U.S. Air Force orders to shoot down foreign aerial discs, at the end of the 1940s."

Howe says she subsequently learned that the rescinding of the shoot-down order was made in the year 1954. Propitiously enough, it was in the year 1954 that an English ufologist anticipated Kissner and Stringfield, by formally accounting for the essentials of their testimony.

Not being positioned to independently verify the data authenticity of Rep. Kissner, I will introduce previously published data indicating how prescient prior authors were, in anticipating the Kissner proposition. And located among the earliest objectives of ufological critique—i.e., military involvement—are detailed data which should dispel certain skeptical assertions that the ET factor is a governmental hoax.

Thus, here is a classic case of both Kissner's principle—a policy of hostility based on a weaponized *fear factor*—and the factual byproduct of the principle. The byproduct is hostility, plus disc-craft evidence. The following passage literally prefigures the Andrew Kissner/Linda M. Howe reportage by exactly 51 years!

How Our "War Of The Worlds" May Have Actually Occurred

From his book entitled *Flying Saucers On The Attack,* 1954 (p. 136), Harold T. Wilkins reports:

"So far it has been found that earth has one weapon which may be, and actually has been—though the facts have been hushed up by the U.S. Air Force—used to cause one type of flying saucer to crash on the ground! *It is radar, which interrupts their drive.* I have seen a letter from an American flier who wrote:

"Several crashes have been reported after a radar beam has been put on the flying saucers. But if you suppose that any man in our armed forces is going to talk about these things, and to tell you what was found when these crashed saucers were examined, you must expect that such a talkative guy is anxious for a court-martial, which he would certainly get, by order of the Pentagon, the headquarters of the U.S. Air Force, in Washington, D.C."

U.S. JET ATTACKS SAUCER!

TRUE STORIES OF THE STRANGE AND THE UNKNOWN

FATE

ANC

MAGAZINE

USAF

April 1957 35c

I HAVE SEEN ZOMBIES

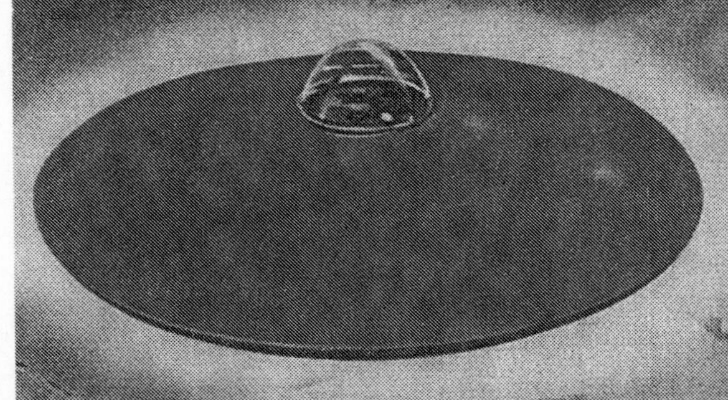

UFO TV SPECIAL EDITION

ROSWELL

The UFO UnCoverup

SEE ACTUAL
UFO CRASH DEBRIS
TESTED AND FOUND TO BE
TRULY EXTRATERRESTRIAL!

3 Time
EBE Award Winner!
People's Choice
Best Picture
Best Music

As a second independent verification of this Air Force disc-shoot-down order, I credit a two part chronology of Human/UFO hostilities. This chronology appeared in the Dec. 1961 and Aug. 1962 issues of *The Searcher* magazine. These two chronologies were both entitled *Are Flying Saucers Hostile?,* by George D. Fawcett. (And Rep. Andrew Kissner confirms they *were* hostile.)

In the first part of Fawcett's hostilities chronicle is the following: "*July 26, 1952*—The U.S. Air Force issued a "shoot 'em down" order to pilots encountering UFOs. This was mentioned in an AF press release." Here, Fawcett verifies Linda M. Howe. She obtained three different newspaper articles reporting this particular July, 1952 Air Force press release. And, lastly—emphatically introducing his 1966 book *Flying Saucers—Serious Business,* veteran journalist Frank Edwards offered this prelude: "Defense Department orders jets to shoot down UFOs which refuse to land when ordered to do so—Washington D.C., *Daily News* 25 July 1952."

George Fawcett's chronicle immediately follows this AF press release citation with a shootdown attempt report: "*1952,* an Air Force F-86 jet pilot opened fire at a hovering UFO in Ohio." If then we consult the 1955 book by the first head of Air Force *Project Blue Book,* we find the details. The very first sentence of Edward J. Ruppelt's *Report On Unidentified Flying Objects* states, "In the summer of 1952 a United States Air Force F-86 jet interceptor shot at a flying saucer." Further details were also described.

George Fawcett's chronology of what were perceived as hostile, or at least negative, encounters number 157 incidents between 1943–1961. Many of these citations are very allied with the perceptions, incidents, and motives today being documented by Andrew Kissner via Linda M. Howe; albeit that theirs is qualified in a context of official offensive military policy. Related military involvement evoked Fawcett's conclusion about UFO secrecy, as of 13 April 1998. The *Enterprise news* of High Point, NC said: "Fawcett believes the government initially kept its findings quiet so the military could keep the technology from its earthly enemies during the Cold War."

On 10 Nov. 1957, eminent investigator Major Donald E. Keyhoe conversed with journalist Frank Edwards. They addressed this issue of "shooting at UFOs." On pages 130–131 of his book, *Flying Saucers: Top Secret,* Keyhoe quotes Edwards, "There's one thing that worries me. Have you heard anything about a new order to fire on the 'saucers?'"

Keyhoe replies, "Only the usual rumors." Edwards explained he had "been told that jets fired on a UFO over Illinois last Wednesday night."

Later, Keyhoe "was reminded of the firing story when I saw Charley Plank at the Civil Aeronautics Administration...Plank stopped me," asking 'What's the idea of this new Air Force statement about shooting at UFOs?'" Keyhoe claimed, "I haven't seen it..."

"You think they're firing on the "saucers" again?" Keyhoe's informed opinion was "... it wouldn't have to be a new order."

"'If there is a new firing order, somebody must be crazy.' Plank paused. 'Or else something's going on we don't even suspect.'" Yes, it is precisely this suspicion, re evidence for a duplicitous militarized agenda, that inspires this writing here. Keyhoe is historically consistent with correlations Linda Howe attributes to Andrew Kissner. Keyhoe is internally consistent with our premise, when quoting his "CAA contact" uttering such a suspicion. And the coincidence indicates how prescient earlier "fair witnesses" were to the UFO/military quandary!

Returning to the Harold T. Wilkins book, it is a veritable clearinghouse of original and historical evidence for the shootdown confirmations, which Linda M. Howe reports on behalf of Rep. Kissner.

Linda Howe reports that the Truman policy of offence toward flying discs was rescinded in 1954, the year Wilkins' book was published. Howe quotes Kissner's determination that the Truman policy was rescinded because the "aerial discs" aggressively retaliated against humans for shooting saucers down. Wilkins' book, *Flying Saucers on the Attack,* p. 271, says, "in the years 1951–1953, *peculiar* and very disturbing *phenomena* have been recorded...although they certainly have not gone without inquiry by the military and scientific branches of the Air Ministries, Air Forces, and security and defense departments.(Italics added).

"These phenomena have...[included] a long series of unexplained accidents to British and U.S. jet planes, besides mysterious disappearances of civilian air-liners, and military planes and personnel.." Wilkins is hereby acknowledging the reported UFO *retaliation* according to Rep. Kissner today—but unknowingly. Re this Linda M. Howe says, "Andrew Kissner reports that within 72 hours [after the 29 May '47] Juarez incident, 29 airplanes crashed around the world killing 198 people...[and] between 5/9–7/31, 1947 more than 600 military and civilian airplanes crashed worldwide.." And yet, one more prescient addendum we shall include—this one by Frank Edwards. In his 1967 book *Flying Saucers—Here and Now!,* Edwards literally anticipates Rep. Kissner's hostilities critique. Edwards says, p. 93, that UFO activity "reached a peak in 1954, and then [seemed] to have tapered off, at least over highly industrialized nations, in late 1955. *By that time shooting at them had also tapered as official policy...*" (Italics added).

Although it is anecdotal, Harold T. Wilkins quotes a passage from a *"Letter from a correspondent in Oklahoma, to the author of this book (July 24, 1952)":* The correspondent, in turn, quotes "a pal in the Air Force" who claimed the military captured an ET survivor: "He comes from another planet, and was one of three others killed in a crash caused by radar, in an Arizona desert, in 1950...[My point in emphasizing this quote is the reference to *radar* as the cause of the crash].

"In this book, I [Wilkins] have spoken of the apparent hostility of some types of flying saucers..." Coincidence? If UFO retaliation is accurate, this would be an ultimate National Security rationale for not volunteering an Official Disclosure about UFOs. First, the *War Of The*

Worlds nature of the 1950s situation is too inconceivable. Secondly, the future spin-potential of the scenario has all the earmarks of another case for militaristic opportunism. We will explore this potential to disadvantage ufology, and disinform global society at large.

Unearthing the Hidden Ground to Mystery and Enigma

Ultimately, Wilkins goes on to cite two detailed chronologies itemizing his "drawing attention to the mysteries of accidents to jet planes...and the unsolved disappearances of both military...and commercial air-liners...Right through 1950 and 1952, in the U.S.A., there were a phenomenal series of accidents to jets and military air transport, and commercial planes." The "phenomenal" number of air accidents Wilkins reported is quite accurate! I have verified this myself. And I strongly urge readers to do their own review of such archival newspaper reportage. For the years 5/9–7/31, 1947, and 1951–53, the number of plane crashes is remarkable. A daily search also reveals many UFO reports juxtaposed with the unprecedented plane incidents(?). One crash citation from 14 June 1947: a Capital Airlines crash near Leesburg, Va., 13 June '47 "brought to 145 the number of persons killed in airline accidents in the past 17 days." The same day, 14 June '47, "A report that the government intends to take immediate and drastic action in an effort to stop airline crashes was circulated today..." Pres. Truman did actually authorize "a Special Board of Inquiry to investigate American air safety." And as Linda Howe reports, one of Truman's appointees to this board was Dr. Jerome C. Hunsaker...also included on the top secret *Majestic Twelve* special studies group. Pres. Truman also sanctioned (*The Mirror* news, Los Angeles, 7/14/52) "Operation Skywatch," where citizens nationwide guarded American skies. They relayed any sightings "to radar stations and ground interceptor commands of the Air Force." (All 6/14/47 quotes from L.A. Herald-Express.)

Harold T. Wilkins also recognized the likelihood that a relationship of stimulus and reaction existed, between UFO hostilities and military-industrial projects. Regarding an incident on July 8, 1947, the discs were attracted to Muroc Airfield, Palmdale, California (i.e., Edwards AFB). Wilkins offered the critique, "It is curious that no plane took off from the ground to try to intercept them—curious, since, as I say, *orders had been issued for that very purpose*." (Italics used emphatically by Wilkins!) Wilkins might not have been surprised to hear "the rest of the story" from Andrew Kissner, 51 years later.

All told, Wilkins performs an unprecedented service. He both prefigures and lays the groundwork for appreciating the Rep. Kissner investigation, with retrospective historical reinforcement. This is an appreciation of numerous mysteries, which are solved by realizing they are the product of human/ET hostilities! A lower magnitude version of the *War Of The Worlds* meme, resurrected for the present day. And Rep. Kissner's study sets a precedent upon which the ufological revisionist historians may need to review and re-vision their own history! One case in point, worthy of such critical thinking, is the "whizzing over the nation's capital" by aerial discs in July, 1952. The period from July 19th into early August saw "a veritable fleet" of UFOs repeatedly intruding into D.C. airspace. There might be a connection between the possible human/UFO hostilities (e.g., retaliation), and provocative appearances where the saucer *"Sky 'Fleet' Flits Over D.C. Again"* (front page banner headline: *The Mirror* news, Los Angeles, 6 Aug. 1952).

Not ironically then, the provocative William Cooper affirmed our hostilities at the MUFON Conference of 2 July, 1989: "Why were so many craft crashing? Because the government was scared. And when they found out the radar was downing the craft—they started aiming the radar at the craft!... and they brought down as many as they could!"

However, the *groundwork* for this post-WWII revisionism does not depend entirely on the contribution by Harold T. Wilkins. As I noted above, the early '60s also saw an attempt to assign critical thinking to human/ET hostilities. The result was the extensive incident-chronology by George Fawcett. Then, in 1967, an updated perspective on this mystery was published by Brad Steiger and Joan Whritenour. Their contribution is entitled *Flying Saucers Are Hostile*.

Resonant Intervals Re-Connect Past & Present with a Conceptual-Continuity of Common Sense Data

From a historical purview, the Wilkins book and the Steiger/Whritenour book both provide continuity in the same way. They both cite *pre*-WWII UFO/ET accounts which provide context for their respective contemporaneous incidents. When taken collectively, these four sources independently support the Kissner perspective on hostilities. Thus, it is fair to say there exists a *pre*-1947 set of data suggesting UFOs acted provocatively toward humans.

There are two scenarios in Steiger's book which are especially relevant to our set of correlations here. First is an attribution to Dr. Wernher von Braun, p. 68. Steiger quotes a comment attributed to von Braun. Wernher von Braun "reportedly" spoke with "a representative of *Neues Europa,*" a German publication, on January 1, 1959. Dr. von Braun is quoted as saying: "We find ourselves faced by powers which are far stronger than we had hitherto assumed, and whose base of operations is at present unknown to us. More I cannot say at present. We are now engaged in entering into closer contact with these powers..." This unverifiable attribution is included because it is plausible, and consistent with an extensive reference to von Braun later in this introduction.

Steiger's other relative point is the following. On page 150, he credits an especially prescient case of "food for thought" as published originally in the July, 1950 *Reader's Digest*. But first—the reputation of the *Reader's Digest* does not represent rants from the lunatic fringe! Thus, in one of his articles entitled, *The "Flying Saucer" Is Good News,* well-known 1940s radio commentator Henry J. Taylor presented a disclosure of sorts. But only now can the public appreciate it for what it revealed. This article, and Henry J. Taylor's biography, is easily verified. I encourage readers to do so.

READER'S DIGEST 7/1950

The "Flying Saucer" Is Good News

Condensed from a radio broadcast

Henry J. Taylor
Internationally known journalist and radio commentator

WITHOUT revealing military secrets, I can assure you that "flying saucers" really do exist. picking up speed, in a lazy sort of way, as airline pilots consistently report, until they move like lightning.

Steiger quotes Taylor as saying, "The flying saucers are a part of a big and expanding experimental project which has been progressing in the United States for nearly three years. It has gone through several phases of development, reaching peaks in public observation in July 1947, January 1948 and April 1950. Meanwhile, the "saucers" have grown bigger with each phase...

"I know what these so-called "flying saucers" are used for. But they are an important military secret. When the U.S. Air Force does see fit to release the information it will be good news—wonderful news—for Americans. Meanwhile, I do not think it is anybody's business to state what they are used for." (This mainstream article coincidentally predated the publication of Frank Scully's 1950 book, *Behind The Flying Saucers,* about the UFO crash at Aztec, New Mexico. Like Scully, Henry J. Taylor was a journalist and a radio commentator, heard coast-to-coast over the ABC network on the program "Your Land and Mine." During WWII Taylor reported via newspapers and radio to America from 32 countries. As an economist, Taylor was a corporate trustee/advisor to banks and companies. In 1957, Taylor was appointed Ambassador to Switzerland. He received the War Dept. citation for conspicuous service in a WWII theatre of combat. Taylor was a Republican. He was a member of the board of directors of The Navy League [New York]. It is these military/corporate affiliations that establish Taylor's credibility for being recruited to "broadcast" a rather shrewd operation of security cover: divert attention by simply disclosing to "your public" exactly what we're doing. "Public incredulity" would settle the rest! See: *Who's Who in America,* 1960–1961, Marquis Pub.)

The political portent inferred by Taylor's *Reader's Digest* disclosure is that of change: a quantum leap in policy, and change of all the primary issues symbolic of 20th Century *reality.* These changes can now be recognized, given all the above. Political maneuvering and strategic (futuristic) policy-making should no longer be holistically analyzed within the exclusive confines of *human* affairs. This means that somehow, an ET-presence exerts an actual influence external to human prerogatives—i.e., an *exopolitical* influence, whereby ETs merely need to be present. Even though, in the experience of most people, ETs represent something tantamount to myth.

The evidence, evermore coherently correlated since Henry J. Taylor authored his *Reader's Digest* disclosure, clearly affirms what Taylor reported: a human manufacture of discoid (or triangular) levitation craft is in fact occurring. *Taylor also says the authorities allowed him to disclose this*—why? Bill Uhouse claimed his security granted him the latitude to disclose, as well. Albeit that Taylor's disclosure was disinforming to the degree that, in 1950, humans may have not perfected disc-craft as far along as Taylor alleged; disinformation being partly true and partly not true. The more common sense version of Taylor's military deployment scheme is featured here by Bill Uhouse. And currently such a military-industrial ambition would be the *hidden ground* to our *black budget* DOD expenditures.

Rarely has it been that specifics about disc-craft development have surfaced in the air industry press. This did occur, however, in the 15 Aug. 1960 edition of *Aviation Week* magazine: *Disk-Shaped Vehicles*

Are Studied For Potential As Orbital Aircraft. The objective was twofold: disc atmospheric aircraft; and disc re-entry craft from space. This article reported that "most of the major air-frame companies have shown considerable interest, and a number of them, including Convair, Boeing and Lockheed, have proposed hypersonic disk-shaped vehicles to USAF." Such disc-craft were reportedly intended to be rocket-powered. But to appreciate the disc-craft account detailed here by Bill Uhouse, it's crucial to connect this *Aviation Week* report with factors conspicuously absent from this *AW* story. In the mid–late 1950s numerous reports appeared in magazines and newspapers about concerted investigations by these same companies into gravity control technology (i.e., electrogravitics propulsion). Then, as with reports like the above, on discs, the gravity control issue went *black*(see *Electrogravitics Systems,* edited by Thomas Valone, www.integrity-research.org).

One brief citation on gravity control is the following. *The Magazine of Space Conquest,* July–August 1958, featured an article entitled, *Has Man Conquered Gravity?:* "Evidence that scientists, aircraft designers, and the Government itself did not consider the control a wild dream...first leaked out in...the *London Daily Express,* on Aug. 29, 1955."

"More than a dozen European scientists," the Express stated, "are signed up by the Glenn L. Martin Company of Baltimore to help crack 'the new frontiers.'"

"The experts will work on problems of space travel and a method of circumventing gravity."

"Then the *New York Herald Tribune* broke with a three-day series on the same subject...the *Tribune* confirmed the Glenn L. Martin Company's secret project. George S. Trimble, described as "a brilliant young scientist" who was head of the new advanced design division of the firm," was quoted as saying, "I know that if Washington decides it will be vital to our national survival to go where we want and to do what we want without having to worry about gravity, we'd find the answer rapidly...that was 1955..."

How to Revise Disinforming History Disguised as a Candid Disclosure

The core purpose of correlating the testimonies here by Melinda Leslie and Bill Uhouse is to verify the credibility of humans simulating ET-technology: an interlock of military collaboration with corporate contractors; thus explained by Col. Philip J. Corso. Of course, the most pragmatic national security byproduct of this reverse-engineering operation is economic innovation—not consciousness raising.

The revelation for today, is that the crashed disc-craft resource, for reverse-engineering ET-technology, is much more logical and understandable. So, how and why did the above assurance by Henry J. Taylor become public? Taylor said, "Our own Navy confirmed...and I am free to tell you...[that] they do exist and they are ours. That's good news for all of us." Well, it's good news for those people persuaded that the alleged experience of Stan Deyo, Bill Uhouse and Bob Lazar is truthful; in that they were employed in the saucer manufacture program. Is it not fortuitous that we have Mr. Taylor's disclosure on the record, telling us what future *"insiders"* would disclose, decades before they later did so? And according to the Bill Uhouse testimony—it would be decades after Henry J. Taylor's statement of record that humans were able to fly a disc of their own construction. In some circles, Henry J. Taylor's disclosure would be termed "effects preceding causes!"

Meanwhile, under whose authority did Taylor reveal that "I know what these so-called "flying saucers" are used for. But they are an important military secret." (?)

Given our possession of magnifying hindsight, for discerning the nuances of ufology today—the *Reader's Digest* affirmations by Taylor are overlaid with candor, suspiciously so! Perhaps the most remarkable factor of Taylor's commentary is the very early timing—1950. Thanks to Col. Philip J. Corso, it is clearly possible that Taylor's statement represents something Corso explained. Taylor seems to qualify for what Corso defined as the chosen model used in controlling ET information: "camouflage through limited disclosure." In 1950, Henry J. Taylor's "limited disclosure" was definitely a camouflaged diversion *away from the extraterrestrial hypothesis*. And Henry J. Taylor's role may have been an exemplar instance of the contradictory or duplicitous agenda as well: In *Flying Saucers: Top Secret* by Major Donald E. Keyhoe, he gives credence to doubts about Taylor. Keyhoe says, page 64, "A statement from Henry J. Taylor, new Ambassador to Switzerland, [says] that the flying saucers were a serious problem—a retraction of his earlier claim, in news broadcasts, that they were secret United States weapons."

Decades of compiled UFO data indicates this control mechanism would translate into layers of *disinformation;* not whole-cloth lies. Truth combined with an admixture of distorted facts. The principle has been effective. As Corso said of the strategy, "and it worked." To wit: as Marshall McLuhan so poignantly exposed the principle, I quote McLuhan from his book on corporate management, *Take Today,* page 92: "Only puny secrets need protection. Big discoveries are protected by public incredulity."(!)

The context for McLuhan's insight is his observation that "Secrecy is a function of slow information movement." And ever since that bellwether year of 1947, info has flowed exponentially more omnifically. How to keep a lid on that? So it was where Col. Corso explains that in UFO matters—"the cover-up is the disclosure and the disclosure is the cover-up." The daily-proofs of this strategy are the memetic words and images of *ET/saucers* increasingly apparent in global culture, as the decades unfold.

Thus, we predictably find in the earliest period of July 1950, an apparent confirmation of Col. Corso. We find well-known journalist of the time, Henry J. Taylor—in mainstream *Reader's Digest*—disclosing "that 'flying saucers' really do exist"; disclosing that "they are ours", or soon will be in the immediate future; and camouflaging the ET-presence by pointing to ourselves. All the while, the skeptical denial of "public incredulity" provokes controversy, and slows the social learning curve of cosmic awareness to a manageable *economic* pace. Shrewd! Taylor even got the jump on Frank Scully's book *Behind The Flying Saucers* (page 3), which Linda M. Howe now recommends for being accurate all along, in light of Andrew Kissner's material. So, our plot—i.e., *learning curve*—thickens! Meanwhile, from *Majestic* document No. 0001—19(24) Sept. 1947, page 16: with ET-crash retrievals "We now have an opportunity to extend our technology beyond the threshold that we have achieved..." (Remainder of passage black-redacted).

Keeping in mind the following pattern of evidence is decades old—from 1960—Donald E. Keyhoe identified and emphasized the ongoing "contradiction" of denying and confirming UFOs! In his book *Flying Saucers: Top Secret,* page 242, Keyhoe was at a loss to make sense of this contradiction. Throughout this book Keyhoe detailed many ways the U.S. Air Force was "still hitting at NICAP and debunking the 'saucers.'" Yet, simultaneously, our Air Force was being credited in the press with acknowledging the *reality* of "saucers"—not debunking them? About this Keyhoe said, "there has been some odd contradictions lately."

We won't duplicate all Keyhoe's encounter cases of contradiction. His entire book is a chronology of cases. The point is that Keyhoe's objective in all his books was to document what he called "the conspiracy of silence" on UFOs. Then, this enigma of disclosure began condensing. After years of directly dealing with Air Force *command and control* of UFO data, Keyhoe had to reconcile policy reversals. How "saucer" information was actually being managed. Keyhoe felt he may have reconciled *the secrecy*—but the contradiction lingered on until this day!

To Keyhoe's credit, as a pioneer, and to his chagrin in terms of what may be called the *exopolitics* of the ET-presence—it would be forty more years before enough data amassed to confirm Keyhoe's original realization!

Example: Keyhoe reports the following case of AF policy contradiction. In an Aug. 24, 1958 press briefing, AF Captain H.C. Cowan "told reporters 'We have to discredit UFO reports just about one hundred percent.' It's clear he was referring to the revised AFR[egulation] 200 order: 'All Air Force activities must reduce the unexplained percentages to the minimum.'" This press statement was identified as being in NICAP files. And the two quotes sum up the official AF *P.R.* position for public consumption.

Then comes the contradiction. And even in this Keyhoe book, he recounts numerous contradictory AF positions. Regarding a 1953 AF "saucer" chase incident: this case was also disclosed in *The Report On Unidentified Flying Objects* by Edward J. Ruppelt. A member of NICAP wrote the Air Force, inquiring about this incident. A written reply was returned by the official AF spokesman on the UFO issue. This official was Major Lawrence J. Tacker. In his reply about the AF chasing saucerian nothingness, Major Tacker did the opposite of what was expected. As Keyhoe writes, "For months [Tacker has] gone all out to prove the 'saucers' aren't real." In his contradiction of policy, Major Tacker "not only confirmed it, he said they had gun-camera pictures and radarscope photos showing the object...and that's not all. They've let several recent Air Force sightings get into the papers." All letters referred to are in NICAP files.

What Keyhoe lacked in 1960 was a greater body of data confirming the contradictory pattern of disclosure: the camouflage through selectively limiting disclosure. This is clear today. But comprehensively compiled indications over time were needed. Evidence now shows the "cover-up" to simply be a *P.R.* ploy, for a much more complex protocol of intentional info management!

For our purposes, as stated, the complexity of ET disinformation hovers over two issues of apparent priority: reverse-engineering, and perceived hostilities. In *Flying Saucers: Top Secret,* Keyhoe was admirably prescient of our focus. Beginning his final chapter Keyhoe gets to the point. Keyhoe had invited journalist Frank Edwards to his office at the National Investigations Committee on Aerial Phenomena. Edwards was a NICAP Board member. Keyhoe said of Edwards, "Like most Board members, he believed there was no proof of [UFO] hostility. In regard to Air Force secrecy, his opinion had partly changed. The main reason for censorship now, he thought, *was to hide intensive efforts to duplicate the 'saucers.'"* (Emphasis added.)

Proceeding onto the escalating question, then, regarding UFO hostilities, Keyhoe offered Edwards some correlations. "It was the first time Edwards had seen all this evidence in such concentrated form." Here, Keyhoe anticipates our own objective, decades later—except, today the correlations have grown exponentially.

From NICAP files, Keyhoe produced an unclassified letter, never publicized. Signed by Major Gen. L.I. Davis, this letter explained an AF order. The order "threatened disciplinary action, even against civilian employees, if there were any more leaks." This AF order was given after a close UFO encounter was publicized by an engineer, James Stokes. Stokes worked at the Air Force Missile Development Center, near Alamogordo, New Mexico. Gen. Davis' letter claimed the order was put out "so that we might get control of a situation which could very easily have gotten out of hand and possibly created a panic..." At the time of this letter, 1957, "For days, all over the globe, UFOs by the hundreds descended into our atmosphere." And the uncontrollable possibility of such a sightings "flap" is a likely *fear factor* regarding secrecy. But does it explain the contradictory convergence of both denial and confirmation of UFO reality?

Major Keyhoe's briefing of Frank Edwards continued on the issues of panic, hostility, and AF plane disappearances. Edwards said, "Why on earth did they put it on record [in NICAP files]?...I know the Air Force used to think the public would stampede at the idea of space visitors. *But people have had years to get used to it."* (Emphasis added.) Perhaps not. Or, perhaps secrecy indicates something else is going on.

Keyhoe responds by quoting an Intelligence report extract: "This Intelligence summary says the captain was convinced they were 'shot at'..." A large AF transport was reported to have encountered an unknown object provoking it. Keyhoe quoted the summary, "The entire crew, including Captain ________, were aware of incidents in which multi-engined Air Force transports have disappeared while flying between..." Hawaii and Japan.

Edwards exclaimed, "Is that the word the Air Force actually used—*'disappeared?'"*

"I'm quoting the exact statement," Keyhoe insisted! Again, this passage is a precedent to that which Linda M. Howe is crediting Rep. Andrew Kissner, in 2005. Thus, in a way, the Kissner data may be ufological *déjà vu,* mainly because ufology seems to have lapsed in its memory of such subtle, but invaluable insights of the past. The existential implications here may actually preclude prosaic, professional skepticism thanks to the complexity alone, of what data now indicates. A complexity of detail like those occurring in any actual setting of circumstances.

All this portends toward actual progress in a *revisioning* of post-WWII history. In other words, the ET factor in this history is an unprecedented catalyst for transforming the perceptual paradigm of earth. Such change demands incubation and sociological midwifery; in our case, media effects. The obvious vehicle for facilitating this trans-perceptuation has been quantum leaps in technology, especially the digital means for speeding up info-movement. And Col. Corso contends that the *techno*-keystone to that rapid progress is rooted in the crashed disc retrievals: a history of alternity, only now might we be prepared to assimilate(?). But, as Frank Scully wisely said in his book: "Not until we, the people...say there are such things as flying saucers, is it authentic." This is the way shifts in the paradigm work.

Before proceeding on, an invaluable calculus can be made linking the above with the following incidents. The forthcoming are some of the most provocative in all ufology. The primary incident is the first-hand experience of a witness named Captain Robert L. Salas, U.S. Air Force, retired. In the eyes of M.I.C. observers, this account would be of greater national security urgency than any disc-shootdown/retaliation data. Although, such shootdown incidents would surely synergize with the reported cases below, of UFOs intruding upon nuclear missile launch preparedness. Our point in emphasis is this: such incidents of UFO interference, would be easily exploited as an ultimate excuse for officially accusing UFOs of hostile threats to national security. Albeit that the Air Force officially declared that, "To date, the flying objects reported have imposed no threat to the security of the United States and its Possessions." The *date* of this official position is *12 August 1954,* as stated in *Air Force Regulation No. 200–2 Intelligence.* An authentic reproduction of this regulation appears to indicate it was authorized by *Lt. Gen. Nathan F. Twining,* (at that time) *Chief of Staff, United States Air Force.* Gen. Twining, as quoted above, is purported to be an *MJ 12* official. The distribution date of AFR 200–2 being 1954 means it coincides with the year Linda M. Howe reports that President Truman's disc-shootdown order was rescinded. Perhaps the insight to this apparent contradiction is: (a) The AF recipients of AFR 200–2 were never in the *MJ 12* "loop." (b) The Truman shootdown order was more *offensive* in purpose. And this distinction may help explain post-1954 disc-shootdown incidents as more defensive. The following would be a reason for defensive action by military minds.

During seven years of service, U.S. Air Force Captain Robert L. Salas served at Malmstrom AFB for about three years. He was a nuclear ICBM launch officer. On 16 March 1967, Cpt. Salas was on duty in the launch control center of the Oscar Flight facility, located adjacent to Malmstrom AFB, one of five launch facilities assigned to his squadron.

Salas received a security call from a guard saying that a UFO shrouded in red illumination was hovering close to the ground, at the base perimeter. Shortly thereafter, the nuclear Minuteman One missiles began to sequentially go "off alert" or "no go"—not in a launchable mode! At this facility 6 to 8 of the ICBMs were shut down.

Given the launch design of the Minuteman One system, such operation losses of missile readiness were impossible. Yet, within minutes of this inexplicable shutdown, Cpt. Salas learned of an identical situation the same morning of 3/16/67. At a sister facility 50–60 miles away, named Echo Flight Squadron, a second "off alert" went down! Echo Flight also reported UFOs and an improbable loss of all 10 of their nuclear missiles. Through a FOIA request, Cpt. Salas obtained a declassified AF Unit History of Echo Flight for 3/16/67. The FOIA document confirmed the mysteriously spontaneous shutdown of the ten ICBMs at the Echo Flight facility. Copies of these original FOIA releases can be read in *Disclosure,* 2001, by Steven M. Greer, M.D., pages 174–175.

There is a thread of continuity here, which links human pursuits of nuclear technology with numerous actions of apparent provocation, as reported, and attributed to UFOs. Obviously M.I.C. personnel would want to know how such alleged acts of provocation were achieved.

Do the Reverse-Engineering Ends Justify a Means of Coercion to Understand It?

In her reporting on Andrew Kissner's insights, Linda M. Howe purposely introduces the reverse-engineering correlations represented by Col. Philip J. Corso. And this, of course, returns us to our contribution offered here: why UFO studies must be perceived as having definitive ramifications regarding covert operations and National Security politics.

In the ufology of today, a trend seems to be condensing. That is the reconciling of data with the political implications. Nearly 60 years of data collection seems sufficient for returning us to what the original set of 1947 facts posed in the first place. However, decades of "discovery process" were necessary so as to dispel skepticism, and prioritize convergent facts.

The facts are, that many threads of continuity align Melinda Leslie's military intelligence intrusion cases with all the above. Yet, this is not well known. It is our purpose here to address such oversight. Although, the big picture will include political circumstances not usually understood to be related...And it is the understanding of how this facet of ufology links the others together that is so greatly needed. In 1993, respected author/"experiencer" Katharina Wilson stated a precedent regarding this area of investigation. In Wilson's book *The Alien Jigsaw,* her Chapter 15 is titled *Government Agencies, the Military and the Aliens.* On page 283, reflecting on her own perceived experience of this human involvement, Wilson says, "This is an issue ufology needs to seriously address, and soon." Now, 13 years later, ufology has not taken heed of Wilson's prescription.

Within the data-set described as ET-abduction—many experiencers report their lives are being encroached upon by Mil-Intel appearing agents. Melinda Leslie and others have documented these cases. Yet, prominent researchers perceive these cases as an anomaly. Therefore, such evidence is deemed irrelevant. This is why the public does not hear of more emphasis placed upon it. This is dubious, since Leslie's focus is on a human activity involving motives and methods humans can relate to. The classification of *anomaly* translates into an inaccurate bias, when authoring a valid hypothesis about the greater ET/UFO issue. And if military-intelligence involvement with ET-experiencers is under-valued, under-reported, and denied sincere analysis, then any "working hypotheses" pro or con regarding "abduction" are thereby faulty!

No hypothesis is valid that does not credibly account for all the data. This elementary *given* will strike academicians as hilariously simplistic. Meanwhile, academicians are those hoping to dispense with "abduction" as *false memory.* This is where *specialization* in thinking and analysis precludes comprehensive hypothesizing.

Dr. John Mack learned this lesson regarding hypothetical specialization, when being briefed on human military-intelligence evidence by Melinda Leslie and Meisha Johnston in 2000. Dr. Mack previously perceived "abduction" as only something "...about aliens taking egg and sperm and traumatizing people." Yet, *Psychology Today* magazine (April 2003) quotes Mack on his evolving perceptions: "I was surprised to find it was an informational thing." Precisely our point here!

What academicians have yet to reconcile is persistent physical evidence for human pursuit of said ET-information, quite possibly done through interrogation/monitoring of certain abductees. Not all abductees, but a substantial number of cases of this are documented. In other words, some of the abductees' *"informational" experience* is of such importance—the evidence indicates humans are apparently using force to secure it. For a list of cases documenting this, contact Ms. Leslie's email: linnie@onebox.com.

Contrary to the sensationalized sperm and egg-count version of ET-abduction, this myopic hypothesis does not explain all the evidence, much less the political implications.

In terms of fundamental validation, it is the human response—not misunderstood ET-activities—which provide a Lowest Common Denominator for framing any ufological understanding today. The LCD prominently identified through 60 years of compiled data is: the Military-Industrial Complex. Not the government per sé. At one time the government was in complicity. But times have changed! Actually, the change crucial to us here, began rather early on. Aforementioned newsman Henry J. Taylor ironically points to it in 1950! Bill Uhouse claims to have spent an entire career at the heart of it: the privatized industrialization of ET-*information*. Uhouse makes sense. Also ironically, Rep. Andrew Kissner provided Linda M. Howe with the institutional framework whereby the principle was forged for privatizing UFOs. Howe quotes Kissner, saying: "Rep. Kissner also was told [by sources including his old "co-worker" who had worked at the Naval Research Lab; and secondly, a "high ranking officer at White Sands

Proving Ground who died in 1998"], that custody of flying discs was permanently vested in a civilian authority, by classified Executive Order in July, of 1948. That civilian authority was the *Atomic Energy Commission* and its National Laboratories. A.E.C. participated with the Department of Defense's *Air Force Systems Command,* the *Naval Research Laboratory,* and the *Research and Development Board* of the Secretary of Defense, established in late July, 1947. All of these [institutions] were established right after the Roswell crashes." In this "civilian authority," a symbiosis was born between our National Lab structure and private civilian corporate contractors. By appreciating this principal relationship, you will learn here, why the Bill Uhouse case couldn't be more crucial in keeping the UFO issue in 21st Century perspective! And a final word on this will be credited to an *MJ 12* doc. entitled 1*st Annual Report, p. 3, IV: 4,* "Because of the unique nature of the material under study, a multi-layered security structure has been in effect. Most of the results have been given to private research and development labs for further study." It is, thus, logical to find a corporate interlock for evidentiary management: a system Bill Uhouse insists is extra-governmentally structured to *govern* the ET-tech replication reality.

The military-industrial circumstance posited above establishes a macrocosm: institutionalizing a human response to evidence for an ET-presence. Embedded therein is persuasive testimony and evidence for the microcosmic response to said human affairs. This includes the witness accounts documented by Melinda Leslie. Her contribution brings logic and systemic connectedness to the technological interface between ETs and humans. Data for both ET abduction and disc crash-retrievals converge in Leslie's documentation. The precedent for such correlations is the historical record of ufology. This record—cited above—supportively antedates the research Linda M. Howe credits to Andrew Kissner. And all this logically suggests national security complicity in disc shootdown/retrieval operations.

The ET-abduction data would also, logically, compel a national security response. This is predictive, whether abduction is largely a psychological or physical phenomena. Security involvement is evident in this data because abduction is a subset of ET-data as a whole. The greater body of ET-data can be fairly defined by the preponderance of military-human response comprising much of it. Generally, ET-data is not limited to being defined by basic abduction accounts. Collectively, the therapeutic abduction-researchers have now investigated thousands of case histories. The nuance of emphasis here is where Melinda Leslie's subset of abduction data converges with the disc crash-retrieval scenario; data previously perceived as unrelated. Increasingly identifying basic abduction cases has led to a greater cohort of Mil-Intel intrusion cases. Why?

Melinda Leslie goes to the crux of the matter in her lectures, regarding humans *re*-abducting "experiencers": "From all the crash-retrievals alone, we know they must be pursuing the technology...Wouldn't they have to be asking, *"'How does this technology work?'"* Thus, Melinda deductively adds, "Wouldn't they [military/corporate intelligence] also be asking, *'Could some of these abductees know something?'"*

Given the national security agenda consistently apparent in all the above—the historical pattern of M.I.C. involvement demands that it be regarded as *the most viable avenue of investigation.* All such corroborating leads should be Exhibit-A priority! Any witness testimonies alleging human involvement in ET-contact cases—as Leslie has discerned—may indicate that "these guys are just doing their jobs." More pragmatically, or more poignantly, Leslie insists "It would be a *dereliction of duty* if they were not involved! That is, if their job is to protect and defend us." And both history and current events validate the credibility of Leslie's perspective; as more and more people struggle with their denial that "creeping militarism" is increasingly obvious in global society. It is precisely this nexus of M.I.C. intelligence imperatives, plus surveillance technology, which make this ufological model more than valid. It makes this model empirically predictive!

One predictive arena wherein this model inescapably applies is security monitoring of the UFO community itself. For the same reason certain ET "experiencers" have their human cases of monitoring and interrogation to volunteer—the community at large should be infiltrated. Melinda observes: "For the same reason, that *if* the abductee's experiences can lead to the monitoring and interrogation of them: Given the covert nature of these reports; given the technology; given the advantages to be gained from knowledge of these areas of evidence, and the economic gains—the M.I.C. guys would have to be interested and involved.

"*Therefore,* they care a lot about the abductees. They are picking them up; they're interrogating them. Not of all, but *some* of them. But if you are one that has that uniqueness to your case...*If* you fit this pattern, you're going to be one of these *re*-abductees...

"For these same reasons, therefore, I can say, *if* we are accurate within the community; if what we're investigating is right; *if* the technology is correct; if the *re*-abduction is really happening—*If* we're correct about even 50% of it—these M.I.C. guys would *have* to be infiltrating us! Because they would have to be on top of: what do we know, when did we know it; who's talking, who's leaking? They would be wanting to spin it in the direction that is useful to them. If we're talking [about] the military-industrial complex, then they would *have to be* spinning it to their economic advantage. And, therefore, they would have to be involved in at least a monitoring capacity. Hence, while there *are Minders* for Mil-Intel intrusion cases, *There Are Minders For Researchers!* There would *have* to be! If we're right, then they are here. And, guess what? We're right. And they're here! [See the Bill Uhouse section] And this is not exceptionally damaging." It goes with the territory.

Disclosing ET History Unofficially Covers Up the Obligation to Take Responsibility for the Effects

Another emphasis of value in this introduction regards official policies of UFO secrecy, and purposes they may serve in our globalized future.

Throughout the current era of UFO studies, a unanimous opinion has been that there exists an official "cover-up" of the best evidence for an alien presence. However, an advantage to our decades of "discovery," is that evidence *now* gives cause to redefine the "cover-up" model. The reasoned essence of such redefinition is the historical issue of *motives* vs. contradictions, comprising the "cover-up" itself, cited below.

In fact, the inadequacy of using a "cover-up" model to define the *official denial* of UFOs is already explained by Col. Corso in his book! Suffice it to say, that evidence for a "cover-up" only exists in the sense that public disclosures about ETs and UFOs are *un*-official. And I am not simply nit-picking over semantics! This is what the evidence available today demands of us to conclude. This understanding is most crucial to appreciating the evidence of Uhouse and Leslie. Both cases give quite affirmative indications that information confirming UFOs *is being disclosed*. Meanwhile, some primary reasoning for why the public should not expect an *official disclosure* is equally indicated. The details will be found in the body of this text.

Regarding the proposition that there is an un-official disclosure plan occurring, we turn again to Bill Uhouse himself. Despite a modest profile, his public disclosures were always qualified. Uhouse claimed they were due to an "agreement" he had with his perpetual security monitors: that he could divulge some of his experience, but not restricted details. Uhouse presents his case in very practical and common sense terms. Thus, with just one un-official discloser, the perception of an iron-clad "cover-up" has changed. Why?

The same holds true for the vetting of veracity and authenticity of the *Majestic 12* documents. Dr. Robert Wood, who is pursuing forensic authentication of the *Majestic Project* documents, reports they are 98% legitimate. If these pages of evidence (some 5,000, of various categories) are mostly accurate—the question now is, why have they been un-officially disclosed? Presuming there is a "cover-up" in effect. Likewise, for ongoing *leaks* of corroborating information?

What are the motives in all this? For, in any investigation, motives make or break prosecution.

Dr. Wood reports that forensic documentarians have shown *MJ 12* files to be largely authentic. If this is credible, what might a skeptical critique of *MJ 12* be? Skeptics would likely conclude that "someone" has contrived the *Majestic Documents;* i.e., they are counterfeit, yet persuasively appearing to be actual intelligence documents. Keep in mind, Dr. Wood claims a few *MJ 12* documents to be original memos, not copies. In any case, such a mass forgery would require extremely nuanced study to pull it off. Meanwhile, such a project of deceit involving voluminous pages puts the onus on the issue of motives even more. Reconciling the *prima facie* rationale of *motive* in this case is, first and foremost, because the *MJ 12* papers exist—whether they are a hoax or legitimate. And of course, from a skeptical bias, the odds are not good that such a deception would be achieved without Intel-archival expertise in the first place: period materials, mastery of Intel-linguistic styles, ideologies, format, etc., etc. This would be likely to require actual Intel-operants as the defrauders—but why? And if the *MJ 12* documents were *perceived* as fraudulent—by association—then the volumes of FBI and military files indicating UFOs are real—these would likewise be falsified. Such an operation implies complexity, involving decades of time, and many people. Thus, this long term program becomes a conspiracy: one of both silence and unspecified logic. It is here where the *motive* factor becomes crucial. The skeptical reasoning here leads to a tautological explanation. But of what? Since, by definition, skeptics are even more anti-conspiracy than they are anti-UFOs! Alas, we still ask, what are the motives?

In our electronic world of media-propaganda, the ultimate question is, which direction of *media-spin* will be placed on ET-information—as it is disclosed, as it is media exposed? As it continues to become voluntarily assimilated into the commons of social awareness, within the global village at large, how is it being used to shape public opinion? And here, we revisit Col. Corso.

In Col. Corso's book, *The Day After Rosewell,* he quotes *MJ 12* member Lt. General Nathan Twining on how the policy of un-official disclosure actually works, p. 75: "What was really needed, Twining suggested, was a method for gathering [UFO] information"... and..."Let skepticism do our work for us until the truth becomes common acceptance."

"'It will be,' General Twining said, *'a case where the cover-up is the disclosure and the disclosure is the cover-up.'"* (Italics added.) Over the past 30 years there has been a mass media consciousness-raising on the ET issue. Basic public awareness *is now* totally commonplace. And at this present stage in our learning-curve, it's irrelevant whether the public believes in or debunks UFO/ET data. The objective is building awareness. Out of awareness is born the ET-presence as an issue. This has been partially based on *leaks,* or *leaked* documents. Why? In other words, for this tactical arrangement to function, two occurrences would probably be observed coincidentally. Our investigations should identify a clear pattern of case contradiction, where sources appear to both make available, and suppress data: i.e., a variety of pathways whereby data surfaces via M.I.C. sources; plus, a variety of means applied in restricting access or denial of data. Especially noteworthy are tactics of witness intimidation (e.g., by phone), property damage, harassment, and professional debunking. All these routinely recur throughout the literature. Not to mention the consensus incredulity of debunking bias in mass media, *to protect the old paradigm*. A reasonable hypothesis can even be made that the George Adamski case itself was a psy-op hoax. It has all the earmarks of being an experiment in the social-acclimation to an ET-presence.

Once again, the candid declaration by Uhouse, that his *un-official* disclosures were allowed by his security monitors, is internally consistent with Gen. Twining's strategy. This is not plausible as a mere coincidence, and the redefinition is not limited to Uhouse. Col. Corso's book was published three years *after* Bill Uhouse began his public speaking. Again, *why?* This question is begged and the answer Corso attributes to Gen. Twining is self-evident in Henry J. Taylor's assertion that: "Our own Navy confirmed...and I am free to tell you [saucers]...do exist and they are ours." The nexus of these convergent facts will prove crucial to the context where Bill Uhouse presents his own account later, below.

An Actual ET Presence is an Incomparable Basis for Contingencies of Futuristic Policymaking

How ET-information may be used to shape future public policy is our final point of introduction. In addressing this possibility, constant attention to our stages of apprehension in appreciating new evidence is instructive. The good news is, that previously disconnected subjects in ufology are now making a lot more sense.

The issue for our future is, *why would* policy-making in an arena as controversial as an ET-presence *be* purposely made public? The answer is, when it will be to the advantage of authorities in doing so—not before.

This is the juncture at which we must diverge from the greater ET issue. Politics, information, and an ET-presence may be converging in a historical perspective now being called *exopolitics.* Within this set are recent developments which, if articulated, may help prevent reactionary contingencies exploiting the ET factor in history. Factual patterns of post-WWII history are a prerequisite for anticipating the future with any degree of accuracy.

Background: During 2000–2001, this author was consulting with Dr. Carol Rosin, as a research assistant to one of the witnesses and legal advisors for Dr. Steven Greer's *Disclosure Project* (http://www.disclosureproject.org). Carol Rosin was also a *Disclosure Project* witness, and political advisor. Her professional contribution to the protocols of public policy-making was important to the *Disclosure Project* as a lobbying group.

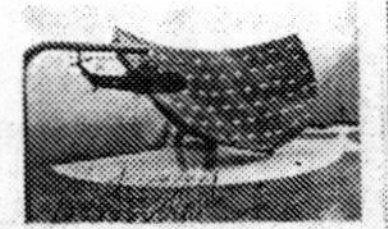
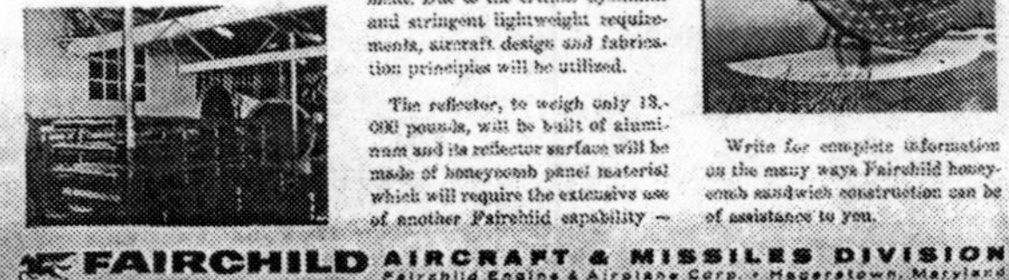

Rosin's expertise is that she was an original legislative lobbyist, attempting to prevent the weaponization of space. Buckminster Fuller gave Rosin the assignment of founding *The Institute for Security and Cooperation in Outer Space.* This was organized for educating the public about U.S. intentions to weaponize space, beginning long before President Reagan announced his *"Star Wars"* phase of the plan. Rosin was the first woman corporate manager of an aerospace company—Fairchild Industries of Germantown, Maryland. Dr. Wernher von Braun was a senior scientist and Vice President of Fairchild at this time. And from 1974–77, Carol Rosin was von Braun's public spokesperson.

History does not manifest by random coincidence! And although von Braun is deceased, Rosin's personal relationship with von Braun brings uncanny clarity to the history unfolding today.

It was fundamental to Rosin's motivation against space-based weapons that Dr. von Braun briefed her on this issue. It was Wernher von Braun's authorial knowledge of aerospace planning which made him aware of space weaponization. Dr. von Braun briefed Rosin on the strategic longevity for how a sequence of U.S. steps would be taken towards the goal. The bottom line is that von Braun insisted the final step would be a tactic of linking ETs with space-weapons policy! But this rather curious pretext would come years after the public is sublimely conditioned for it; for acquiescing to it.

Carol Rosin was inspired to organize a movement against space-weaponization by Dr. von Braun. He repeatedly claimed there was a preconceived sequence of threat options contriving the long range policy: these comprised the American strategy for insuring military superiority. Plus, the necessarily astronomical D.O.D. budgets for sustaining such a technological futurism. There were no documents available then to corroborate von Braun as correct. However, close scrutiny of both official and un-official documents available now can be recognized as so highly consistent as to preclude coincidence. The strategy Rosin insists she learned from von Braun in 1974—thus far—has been entirely correct. This, I mean generally and procedurally, mainly because a fair rendering of space weapons history indicates that a succession of contingent *realities* must surface to demand the program many scientists insist is of improbable value, and wasteful of resources.

So, the question is not whether the Carol Rosin account is true. Rather, is it possible that evidence can be brought to bear consistently lending credibility to her or von Braun's proposition? Fortuitously—many independent sources of evidence for this pattern exist.

First, Wernher von Braun's assertion: Carol Rosin quotes von Braun describing this on several occasions. It was the last part of a greater scenario that von Braun emphasized to Rosin as involving a media campaign. These were ways which the covert operation was going to be educating the public; to scare them. For, the final stage was going to be ETs as a threat. And what von Braun was really concerned about was just such an ET disclosure, plus psychotronics. This Rosin/von Braun account is factored into our overall critique here because their purpose was to expose a pattern of history-making, to be contrived by a succession of officials; because this alleged contrivance has been predictive of the mounting evidence for deceptive causes of the War on Terror in 2005; and because the Rosin/von Braun account could thus be one reason for the ET "cover-up"-disclosure. Worthy of noting also, is that Carol Rosin was publicly divulging the sequence of contrivances three years prior to George W. Bush declaring the War on Terror in full swing.

Back in the early 1970s, when Rosin first met von Braun, he expressed concern for the entire scenario. One of the punch lines was the deliberate way this game of weaponizing space was going to be played: i.e., the goal is *funding* the implementation with successive rationalizations.

Threat 1—The Soviet Union was the enemy against which we had to build space-based weapons. In receiving her 1974 briefings from von Braun, Rosin understood clearly. The space-weapons threat-strategy meant a near term *anticipation* of an end to the Cold War. We provide independently discerned evidence for the prescient nature of this "anticipation" at the end of this introduction. And an invaluable fact consistent with this threat is WWII revisionism now, by mainstream academics: that nuclear bombing Japan was not exclusively done to end WWII; it was done to intimidate the Russians, *as the first act of the Cold War!*

Threat 2—This would be global terrorism.

As we all know, our state of the world today engenders constant *en guerre* alert to this threat. Establishing *terror* as a force threatening global stability gives *terrorism even greater distinction. Terror*-ism thus, coincidentally, supplants Communism as the new global organizing principle. Our D.O.D. budgets are now much higher than even during the Cold War.

On 24 Sept. 2005, growing masses of people demonstrated their realization that our second Gulf War in Iraq against terrorism was entirely rationalized with an invalid purpose. And according to Carol Rosin, the current war on terrorism in Iraq began many years prior to the First Gulf War in Iraq.

In her position as Corporate Manager of Fairchild Industries from 1974–1977, Rosin explained what motivated her resignation. This resulted from her attending a strategic planning meeting in 1977. This meeting was held at Fairchild Industries, in their "War Room." The people in attendance at this conference were strategic consultants, industry representatives, and "revolving door" military/intelligence personnel.

The planning was about potential *terrorists* and leaders like Saddam Hussein, who were not enemies at that time. Rosin reports that the objective was the intent to antagonize these potential enemies, plus the anticipation of a war in the Persian Gulf—a Gulf War! But this was 1977! The crucial point in this contingency, was that creating the first Gulf War had to coincide with the period when there would be $25 billion allocated to the space-based weapons program. They were planning ahead for the *need* of new weapons systems, and then filling that need. And as Carol Rosin reports, the First Gulf War happened exactly as it was planned, and on time! (See *Disclosure,* pages 255–261, with testimonies of 67 military/industry witnesses, by Dr. Steven M. Greer, 2001.)

Threat 3—This, currently, is a second order backup threat to terrorism. The non-aligned countries called *rogue states,* etc., developing nuclear or other Weapons of Mass Destruction. These WMD-states are being simultaneously prepped as further reasons for space-weapons in 2005, e.g., North Korea and Iran.

Threat 4—A not too distant future threat of asteroids or comets from space. The most recent media-event hint for moving public awareness in this direction is the NASA *Deep Impact* project. Reported 3 July 2005, *L.A. Daily News,* this space probe was designed to intercept and collide with a flyby comet.

However, on 24 Feb. 2004, the *Los Angeles Times* afforded much greater credence to the cosmic fear of celestial collisions. An article titled *Scientists Gather to Discuss Ways to Block Disastrous Asteroid* reported on the first major conference of its kind on this threat. A primary strategy prescribed to preempt this threat is "deflecting asteroids with nuclear warheads." The conference was called "The Planetary Defense Conference: Protecting Earth From Asteroids."

Congressman Dana Rohrabacher (R–Calif.) "has introduced two bills" supporting "research on threats from outer space." Rep. Rohrabacher spoke at this conference saying, "(Osama) Bin Laden was out there like a near-Earth object for a long time...it took 9/11...for us to pay attention to that threat. I hope it won't take that long for us to recognize the threat of near-Earth objects." The U.S. government has charted asteroid flight paths since 1998.

Threat 5—And the final, ultimate threat requiring our American weapons-control of space is extraterrestrials. This threat would make sense to a public largely motivated by *conditioned fear factor,* reinforced by years of recurrent scenarios like those fostered in 2005, plus equivocal or ambiguous ET infotainment. Coincident with the time of this writing were two new exhibits from the *Aliens intrude upon earth* genera of TV-agitprop: *Threshold* and *Invasion.* These TV-series began in Sept., 2005. Are these kind of shows *Tracts for the Times,* the paradigmatic end times? Taking Col. Corso's media distraction *"strategy"* as a benchmark—could it be these media tracts resonate or mirror a sentiment of ET-threats, socially projected? Has society reached a *"Threshold"* of perceived *"Invasion"* into the provincial-paradigm of humanity? An intentional mood, i.e., a Cold War with ETs, based on human/ET hostilities alleged to have occurred in the 40s and 50s?

Carol Rosin's job, as von Braun taught her from 1974–77, was to learn about this whole situation, doing whatever job description she had in the industry. This began by lobbying against conventional weapons. But Rosin emphasized that the weapons part of the strategic planning was not von Braun's real concern, he claimed. It was actually the extraterrestrial scenario—the last threat in the sequence—that may be pulled out as a big threat.

As a consultant to Rosin, I specifically asked her whether she felt von Braun's concern about an ET-threat may have been based on

Executive knowledge that there were ETs? In other words, that the national security planners were ultimately going to *spin control* the basic data about ETs in their direction? So they could continually reinvent themselves, as they anticipated the end of the Cold War and beyond? Basically, that is exactly what Wernher von Braun said, regarding the latter part of this long range plan—a plan showing signs of coherency during these same early 1970s, in Rand Corporation feasibility studies, which began as *Project Seesaw,* then *"Star Wars"*/SDI, now BMDO or a Global Protection System. (See *Paranoia mag. Issue 27, Fall 2001*). And according to Col. Philip J. Corso, there were "scores of other camouflage projects...and even the Space Defense Initiative...had something to do with alien technology."

Over those next few years, Carol Rosin was confident that Wernher von Braun was "implying" to her, and did so, that he personally knew something about the ET issue. He never told her what it was. And Rosin assured me she felt "devastated" now, that von Braun never compromised his obvious security obligations on the matter.

We know from the history of American rocket development, however, that Wernher von Braun was directing V-2 tests for *Operation Bumper,* at White Sands Proving Ground in 1947. In Mercury 7-astronaut Gordon Cooper's autobiography, *Leap of Faith,* he discloses an account of von Braun witnessing two UFOs pacing a V-2 rocket test, 10 July 1949. And now that the "leaked" *Majestic Documents* are being authenticated, we have a possible insight into what von Braun actually may have participated in.

As stated above, Dr. Robert Wood is confident that he has succeeded in forensic authentication of *MJ 12*. Therefore, the evidence for von Braun's probable access to knowing what he confided Carol Rosin with, is located in the following document: Air Accident Report on "Flying Disc" aircraft near the White Sands Proving Ground, New Mexico (16 July 47), pages 21–26 of *The Majestic Documents,* edited by Dr. Robert M. Wood and Ryan S. Wood, 1998. Perhaps, this document explains how it came to pass that von Braun possessed the knowledge he reported to Rosin, because this Air Accident Report names Dr. Wernher von Braun as a "Flying Disc" crash retrieval technical advisor. It is also a corroborative detail that the *MJ 12* documents should anticipate our current space weaponization process! The following passage is from a document entitled *Relationships with Inhabitants of Celestial Bodies,* Draft, June 1947, p. 5. It states, "Military strategists foresee the use of space craft with nuclear warheads as the ultimate weapon of war. Even the deployment of artificial satellites for intelligence gathering and target selection is not far off." And lastly, in his book *The Day After Roswell,* Col. Corso quotes himself making a recommendation to secure Wernher von Braun's participation. Corso tells his superior, Gen. Trudeau, he would "go to scientists with clearance we can trust [like] von Braun for advice," on the potentials for tech replication, based upon "alien technology."

There is one more citation worthy of supplementing the above paragraph of correlations. It is a reinforcement, a report of historical congruency, where many of the same details converge.

From the July–August issue of *Flying Saucers* magazine, 1958, a fortuitous article appeared: *The War For Outer Space,* "The race to the moon...is coming into focus as a race for control of outer space."

This article begins:

"Why the great urgency to dominate the space beyond the earth's atmosphere? United States Senator Lyndon B. Johnson recently explained: "Whoever gains the ultimate position gains control, total control, over earth..."

"So it is not mere scientific interest in outer space that prompts all the talk about manned satellites and moon bases...now it is a race for control of the earth through control of outer space...There is a tone of urgency whenever military leaders discuss the matter."

Today, 48 years later, the genesis of the concept phase has matured into a merging of U.S. Space Command into U.S. Strategic Command.

And the intelligentsia of conscious activism can't stop the momentum that weaponizing space has attained. The article goes on to quote numerous leading rocket/space authorities—but mostly, Dr. Wernher von Braun. In this article von Braun confirms, consistently, the first rationalization for weaponizing space, that Carol Rosin attributes to von Braun—the Soviet threat.

The *Flying Saucers* article quoted the book *Rockets Beyond the Earth,* by Martin Caidin, 1952, in which a definitive question was posed to von Braun. "How soon do the nations expect to put up a manned space station...? 'That is a question of the financial effort behind it,' explains Dr. von Braun. 'That is the root of the whole question'"(!) Yes, both then and today the scientific worthiness of the space station has proven to be unclear.

So, once again, the critical thinking applied by the "flying saucer" community presaged society at large. And the more data connections ufology makes, the greater the clarity that a hidden ground to militarizing space is the ET presence itself. Indeed, *Flying Saucers* still retains the last word, by concluding with: "We need no space-fiction thriller or television program to presage the future for outer space if one nation could control it."

The Hidden Ground to Politics of Economic Militarism would be Exopolitics

Now, onto the homespun American creation of a need for threats—then filling the need with a lucrative business of weaponizing outer space. About a year prior to the George W. Bush Administration taking office, a document definitive to the above was published. It was entitled, *Rebuilding America's Defense,* published by The Project for the New American Century group. This position paper sought to secure continually escalating D.O.D. budgets allocated for the technological transformation of the U.S. Military. It is a doctrine designed for *creating vacuums into which their ideology can move* in the guise of F.E.A.R.: False Evidence Appearing Real.

The centerpiece of this high-tech transformation had been the U.S. Space Command project to weaponize, and thereby dominate, space. As with Pres. Reagan's originating *"Star Wars"* phase of this program, the U.S. Space Command incarnation was the current status of the agenda. The U.S.S.C. charter coordinates with the serialization of threats—*the need*—in von Braun's briefing. All of which demand a space-weapons solution (e.g., front page New York Times, May 18, 2005: a national security directive for updating space policy including space weaponization.) The reason why the USSC *was* commandeering American space dominance is because in October, 2002, policy shifted. The USSC mission was merged into U.S. Strategic Command (Stratcom)—see *Aviation Week,* 4 April 2005.

Historically exemplifying America's principal intent to militarize space is the following disclosure by a major contractor—Northrop. In their corporate magazine, *The Northrop-Norair News* of 28 Sept. 1960, they reported on their space weapons plans. Their Norair Astro division developed "maneuverable satellites"—*"Astro Sciences* has conducted numerous feasibility studies on earth orbital, lunar and interplanetary strategic weapons systems."

Consistent with Northrop, the 15 Aug. 1960 issue of *Aviation Week* magazine reported on the predictive "usefulness of sending man in space." The Space Science Board of the National Academy of Sciences—National Research Council "paradoxically" concluded that "*the only real role* man will have in space will be of a military nature." (Italics added.) Today, the U.S. Strategic Command institutionalizes these original values. They are an instrument of strategic doctrine requiring space weaponization. So, what's going on here? Is *space* the *Final Frontier* of humanity's quest for science and existential knowledge—or not?!

This ongoing strategy of creating needs (threats), and filling those needs with big budget solutions is actually quite pragmatic. It's also the means for throwing policy critics off guard by constantly *creating new realities.* Authoring new realities is a principle for keeping the

ET-presence in perspective, as an indicator of an alternate reality, a transcendent reality that includes *them.* *Pulitzer-Prize* winning reporter Ron Susskind was challenged with the tactic of creating new realities when researching his reports about W. Bush Administration officials.

By writing personality profiles about W. Bush officials, Susskind came to reconcile his being briefed about this *modus operandi,* whereby the administration achieves its goals. In an *Esquire* magazine piece, Susskind anonymously quoted a "senior advisor" to Pres. W. Bush. This "advisor" explained the insight of creating new realities, which goes to the heart of the Presidency, and its agenda of contingencies. The "advisor" told Susskind that the journalist represented "the *reality* based community." And such people believe that problem solving emerges from their study of *"discernible reality."* But the "advisor" defined an alternative scenario being played out in our 21st century era. Susskind was told, "The world doesn't really work that way any more...when we act, we create our own reality, a new reality."

Ron Susskind interpreted this to mean creating the current events of the administration's agenda: the *product* manufactured by our dominant culture of orchestrated influence in the world; on the stage of global theatre. Or, as Marshall McLuhan foretold in his book *Take Today,* "The familiar idea of 'making the news' now yields to making the world itself." Profound, though this may be—most folks acquiesce to having been taught that *reality* is constant, the same for everyone: consensus. Nothing could be further from what actually is: *perception* is reality!

The W. Bush Administration has proven to be overtly ambitious in advocating its orchestrated culture of "assertion." As Susskind reports it, "this *idea* of an ability to alter reality through confidence and certainty, and a kind of tactical forcefulness, is right at the heart of the administration." Thus, the administration *programs* "an ethos that believes forceful action, in whatever direction, is in its way its own justification."

"Essentially," the W. Bush "advisor" said, "you [Susskind] will study what we do. And while you're studying that new reality, we'll act and create another one...We are historied actors, and you will be left to study what we do." (Ron Susskind reported this on the *To The Point* show, KCRW radio, 29 Oct. 2004, Los Angeles). All this is a strategy of maneuvering opportunities "to refashion the world," as Sec. of Defense Donald Rumsfeld is quoted as saying. This goal of reality creation is clarified in other publicly available documents (all consistent with von Braun):

In Feb. 1997, the U.S. Space Command published *Vision for 2020.* In its mission statement, USSC said, "U.S. Space Command [is] dominating the space dimension of military operations to protect U.S. interests and investments"; yet, a low priority was set on using space domination for that great contemporary theme of "democracy promotion." *Vision for 2020* also said, "The globalization of the world economy will continue with a widening between the haves and have nots." Such chronic inequity thereby morphs into an incentive for weaponizing space; for marshalling countries, politically exploited by such inequities, with the power of "full spectrum dominance." As opposed to a humanitarian redistribution of wealth. But why are the space weapons so necessary?

Ironically, China has led an international effort at the U.N. to *preserve space* for peaceful purposes. These international intentions have been consistently blocked by U.S. missile-defense advocates since the Clinton era. As scholar Noam Chomsky has stated, this American rejection of reserving space for peace has not been noticeably reported, "though of extreme importance." Yet, technological evolution has merged with the consensus paradigm of adversarial-mindedness to make *"MILSPACE"* inevitable. "Space" is the post-modern medium for commanding operations of info networks, weapons, intelligence, surveillance and reconnaissance: the cosmic security complex.

The W. Bush Administration is now expediting space weaponization. Subsequent to the announcement of Bush's *National Security Strategy,* Sept. 2002, the means for implementing USSC strategy required a position statement by USSC. Here the USSC asserted a redefined goal for its militarized space policy: the stated policy during the Clinton Administration was "control of space" for military purposes. The *redefined* W. Bush policy is the "ownership of space" for military purposes! This shift in goal orientation would seem suspiciously necessary, as a preparation for the threatening ET-contingency that Carol Rosin quotes Wernher von Braun as exposing. It makes sense. History does not occur through random selection. And it would make sense on the level of previous policy. The research Linda M. Howe credits to Rep. Andrew Kissner may have formally established a historical precedence of perceiving ETs as a threat. And yet, what would be the viability of such a threat? Is it contingent on the success rate of historically contriving the prelude of threats we are seeing, and the global *fear factor* provoked by them? E.g., North Korea and Iran...

In a lecture delivered to the International Relations Center, New Mexico, Jan. 2005, Noam Chomsky emphasized the U.S. Air Force's Space Command. Chomsky quoted the purpose of USSC space weaponization as, "instant engagement anywhere, with highly lethal offensive weapons, which can strike anywhere on earth without warning.

"The whole world is under surveillance by sophisticated satellite...systems. This also was, as far as I know, not reported at all. Certainly not much!"

We know from the current war in Iraq that policies of our Dept. of *Defense* have flipped into being offensive and preemptive. So, we can fairly deduce that von Braun's prefigured threat sequence is not, thus far, a bluff. The following indicates how coherently contrived it is.

Enter the Momentum of Conflict Continuity

Actually, the basic motive postulated in Wernher von Braun's briefing to Carol Rosin would be "the end justifies the means"; the corollary to which is "creating needs and filling them"; not to mention "self-fulfilling prophesy," and "might makes right."

Mainstream journalist Robert Perry documents this principle, whereby *realities* of deception are created; the punitive fulfilling of arbitrary needs *are achieved*. Perry's researched account helps verify the prosaic credibility level of Rosin's attribution to von Braun. Here, the premeditated pattern continues with clues pointing both forward and backward in time (history).

In a radio interview (2 Jan. 2005, KPFK, Los Angeles) about Perry's book, *Secrecy and Privilege* (2004), he was questioned about Pres. W. Bush's agenda implementation: "There seems to be a formula as to how Bush does things. And that is, to create a sense of crisis, a fake crisis. And then to, through fear, stampede the American people into a certain direction. That, often, is into the waiting arms of his [Bush's] corporate allies. Is that a fair assessment...and can we expect more of that in the second term?"

Perry replied, "I think you're right. That has been, certainly, the way he's approached the post-911 period. And the whole Iraq War was to create this sense of much graver danger than probably existed.

"One of the elements of *Secrecy and Privilege,* my book, was to look at how that *(formula)* even goes back to the late and mid-70s when [W. Bush's] father was head of the Central Intelligence Agency, in 1976. The idea of the conservatives at that point was to exaggerate the danger we faced from the Soviet Union.

"Even more recently, I had this interesting [discussion] with a former senior CIA officer. [He] was telling me that they were having very detailed information from...their top agents inside the Soviet Bloc: describing the coming collapse of the Soviet Union by the mid-70s. *And this was not welcome information! You might think it would be. You would think that the collapse of [America's] Cold War enemy would be good news? But it was not good news!* (Italics added.)

"And the reason it wasn't, was because, to justify continued military spending at the high level; and also, other interventions around the world...There needed to be a heightened fear from the American people about the Soviet Union. So the decision went. And this begins in 1976, with the so-called *Team B*—who were a group I would call *neo-cons,* like Paul Wolfowitz—that are allowed into the CIA, to basically chastise the analytical division there for not being alarmist enough(!), about the Soviet Union.

"You would then begin seeing the building up of this huge *straw-man:* of how the Soviet Union was getting more and more powerful and was creating this huge 'missile-gap', the 'window of vulnerability' [etc.]...and the reality was quite the opposite! The Soviet Union was crumbling. It was having trouble keeping its own satellites in line, let alone trying to conquer the United States. But, again—the fear, the exaggerated fear that was pumped into our own system justified an increasing escalation of the Cold War in the 1980s. When, in fact, many people even like Nixon and Kissinger [claimed] *detente* was a real possibility by the mid-70s. *Instead, we went the other way*(?)!

"Then, when the Soviet Union—this giant *strawman*—fell over around 1990, ironically, the people that exaggerated, that made up the *strawman...*" those neo-cons are now resurrected with W. Bush. They are clearly and deliberately *real*-izing new *strawmen,* like Iraq and its 911 threat of WMD, that didn't exist: Weapons of Mass Distraction.

So, with Robert Perry, we have a totally independent researcher providing a prosaic, but definitive background to the von Braun-Carol Rosin briefing. Perry unintentionally fills in corroborating facts, and historical credibility to this strategic premise of von Braun's threat-sequence briefing. Plus, Robert Perry emphasizes why U.S. Foreign Policy "went the other way" precisely at the time (mid-70s) Wernher von Braun is credited with anticipating where "the other way" did, in fact, lead us: past "the end of history" as the Cold War deconstructed, and onto a new Pearl Harbor (9-11) of terrorism and rogue-state WMD! What a coincidence...And the same strategically deceptive creation of threatening realities occurred in the Gulf of Tonkin incident, that led to escalating the Viet Nam War. These are the "exemplary actions" taken by our doctrine of conflict, but these "credible pretexts" for the conflicts prove dubious at best. Because the history of these "actions" did not occur by coincidence.

This post-WWII pattern of history has been forged. It is goal-oriented towards military domination of space. Robert Perry, however, is not the only investigator or scholar now revisioning history with this pattern. Author/historian Gore Vidal can serve as a second clinical opinion regarding the precedence of threat creation.

In a radio appearance (13 June 2005, Los Angeles), Gore Vidal was asked why the American electorate seems to relegate their government with the role of security blanket. Electing it to office for instituting creeds and mores, something the government is not designed to provide? Vidal replied, "...The answer there was...codified by Pres. Harry Truman who started the National Security State. And [Truman] was warned that if he wanted to raise all this money to fight communism, to get [the necessary political] support, Senator Vandenberg, a Republican, said: 'well, if you want to have this military buildup in a world at peace'—this is 1945–50, we were on top of the world. Nobody endangered us!—But, Vandenberg said, 'if you want this military buildup when there is no enemy in sight...you've gotta scare the American people to death'!

"We've had nothing but being frightened to death! Now, we're being frightened by *terror*. You know, it's an abstract noun! Nobody knows what it is, who it is. We certainly know when we get hit by it. But then we have to sit down and think, why did we get hit? What have we done that has led up to being hit?

"So I would say [looking back], there was Franklin Roosevelt, *our Augustus,* our great President, and he said 'We have nothing to fear, but fear itself'. He starts off with that, saying don't let them frighten you into wars; don't let them frighten you into giving all your savings...

"Then comes Harry Truman [saying], 'We're in a world today, in which communism is on the march'...And suddenly the difference between two presidents—*and it's been the official line...ever since: to frighten us with terrible, terrible enemies.* That's how we've got such a polity that prefers to talk about 'don't burn the flag; In God We Trust'—all sorts of slogans—and not *do anything* about containing the military, which is hemorrhaging us of money." (Italics added.)

And the pattern feeds upon fear: fear of *the bomb, fear of the Reds, fear of Osama bin Laden, fear of "worlds in collision"*—and ultimately perhaps, *cosmic fear* of alien invaders. Given the prosaic but clearly deceptive political realities recent history faces us with—if you have studied the evidence for the ET presence, the probability of ET being disclosed as a threat is predictably the only purpose for there being any disclosure. And certain skeptics, rationalists, who deem the ET factor a psy-op myth—they anticipate an ET-threat as well. The new information mentioned above, about the credibility of human/UFO hostilities, could be an ideal ploy for *disclosing* ETs are real and threatening. Time will tell.

So, to reinforce the instrumental correlation in focus here—there does exist a statecraft purpose for shape-shifting realities, before our very eyes. In one additional exposition on the phenomena, credit goes to Prof. Robert J. Lifton (Harvard), speaking at Amherst College, Mass., Nov. 2004.

Prof. Lifton's discernment of the system for this principle is as follows. Lifton pursued a study of "apocalyptic" conflict. He found he had to reconcile the reactionary methods used in the post-911 crisis we are now immersed in, mainly because this real-time fear factor meant recent post-911 conflicts were "a little different than" the historical cases Lifton researched. He found that American "leaders have systematically manipulated" 911 by exploiting fear "for their own political and military purposes."

Such manipulation is not itself unique, historically, but Lifton "came to realize" that certain post-911 methods of *apocalyptic* conflict opportunism do have a unique psy-op dimension: i.e., there was a "quest for the ownership of the mind" involved. The post-911 American apocalypsis was for "control of political life," and actually the "ownership of reality" itself. This new twist in the strategy of "thought reform," you might say, is a *new age* approach to our post-Cold War threat of today. It's new age in the sense that *reality* is flux, not constant. Now, personalized *thought control,* "brainwashing," "ownership of the mind" or *Mind Control* gives way to what may be a more occult insight into techniques of social engineering: the mass psychology of consensus-*reality* itself. The new age *precept* here is that the reality you live is the reality you are responsible for, both consciously and unconsciously. The unconscious aspect can be vulnerable to manipulation. It is therefore a wise choice to take the suggestion of an old bumper-sticker adage: "Question reality"!

We should question it, because today, the perceptual-centrism of imprinting upon Americans the perspectives they learn, is almost entirely molded by the *effects* of mixed-corporate media. These effects—sensorial and psychic—are separate from the media content. But a synergy occurs between them. The doubtless fact that our ubiquitous media is conglomerate controlled, explains the obvious purpose whereby media molds its content. The user is overwhelmed with *everything happening* in the world all at once, and consuming it *all at once.* By *spinning* the viewpoints of its presentations, media imprints conformity to ideological doctrines of value. Through this equation of sensorial media effects, plus *spin*-bias of broadcast style—*voilá,* you have a mass perceptual alternity exploited by censorship by omission. People will believe in something that is emotionally presented, and repeated long enough. Once they believe, then they perceive it as *reality.* They conform their sense of identity to media trance-projections. This happens even though this sound bite rendering of current affairs is only partially or *virtually* the actual case. This formula insures the social conditioning. This system of applied media establishes virtual-realities which then can morph into ahistorical myths. And lastly, as Gore Vidal has observed, American corpocracy has been the most clever in all of history, in persuading people to *"think* against their own best interests"...a principle crucial to our virtualized ways of life.

Of this principle Prof. Lifton says, "one can look no further than the *reality reversals* of the W. Bush Administration at the present. Really looking at Iraq, to start with and to finish with.

"In terms of *reversing* Reality—when things are absolute chaos, as they are now, [the war] is being presented as a success story...and similar things are done with domestic issues.

"This kind of *reversal* of Reality is more than just [media] control of spin. It's, again, an effort at *ownership of the mind,* or of *reality control!"* Philosophically, such manipulation is explained in the phrase *ordo ab chao:* order out of chaos. The *Newspeak* of *1984*.

With Prof. Lifton's analysis, we once again have a qualified observer deconstructing that code of contention which provokes so many crises in our high-tech era. As Ron Susskind put it above—"This *idea* of an ability to alter reality" clashes with beliefs in an immutable reality, or set or perceptions. This is a sign of our times: *what is* reality, or the paradigm, or the world-view of any given population? And the ultimate mystique which vitalizes *reality* as a supreme meme of change, is the awareness that realities are actual, and realities are virtual. You may acclimate to them by default, or you may be persuaded by design (conspiracy).

Meanwhile, many are the reasons for interlocking goals and motives, which this historical pattern of serialized-threat *(reality)* creation would be designed to achieve simultaneously. Yet, the technological hidden ground to all this is only sincerely perceived by the UFO community. This is because, in the pursuit of ET studies, the UFO community is not holding onto the old paradigm—they are helping to forge the new one.

The bottom line for all this has two facets: the paradigm of technological dominance; and, the economy. Not the disruptive concept of an ET presence per sé. It must be more subtle and pragmatic than that. Remember, WWII was the economic panacea to the Great Depression.

This *bottom line* was clearly specified to Linda Moulton Howe by Col. Philip J. Corso. In her 24 April 2005 lecture, San Diego, California, Howe reported the obvious, according to Corso's "need to know."

Linda M. Howe described her two private discussions with Col. Corso. Howe queried the technological mystery for performing the cattle mutilation phenomena. Corso confirmed a "need to know" answer to this. He claimed to have read reports of UFO-crash retrieval teams finding ET-technologies apparently used in cow-tissue sample collection. Apparently, an instance of this was the *UFO Crash at Aztec* (see the book on this by William S. Steinman).

Howe quoted Col. Corso's personal response to her regarding the issue of retrieved technologies: "'...The advanced technologies needed to be kept out of the hands of human enemies, and developed and patented by American companies.' That, he said, was the 'highest priority!'" Based on how our system functions, you know this statement is plausible. It makes sense, we can appreciate it, and it would be at the top of any list as to why there would be no Official Disclosure: i.e., the bottom line is the economics of cultural momentum, where ET is philosophically or existentially threatening to a consumerist *reality.* Yet, ET is also the resource model for instant quantum leaps in developmental superiority. Economic proponents of our prevailing "order," will not allow a shakedown of this paradigm. It's bad for business.

Pro or Con?—Media is Our Learning Curve Provider

A policy of *non*-disclosure and a "cover-up" are not the same thing. We still must reconcile the formidable evidence for Col. Corso's "camouflage through limited disclosure." In other words, there is media proliferation of a *symbolic* ET-presence. Society now assimilates a cultural saturation of ET imagery, ET icons, ET *symbology* and ET/UFOs as memes of sociological and technological futurity. This is clearly an ongoing media campaign, if one is confident that ET evidence is factual. Most of the time ETs are the antagonistic *other,* in films. This is propaganda. But your degree of skepticism determines how you are *imprinted* with the content. I.e., *"The Strategy,"* as Corso euphemized it, is a protocol of diversity. The material evidence for ET "we denied," Corso explained—"while encouraging science fiction writers to make movies." "We" did this "...to blow off some of the pressure concerning the truth about flying discs...and it worked."

An anthropologist would say, *masses of people in society are subject to manipulation by symbols and stereotypes managed by individuals operating through their control of communication media.* If the "stereotype" of ETs is variously threatening—which they are in films, etc.—then the conditioned response by humans may be dubious, but emotionally imprinting. Thus, the symbolic or virtual ET-presence has been established via subliminal advertising or subliminal intervention of direct marketing. This *Trojan Horse* is uniquely true with ET/UFO memes featured in TV or radio commercial ads. As Col. Corso quotes Gen. Nathan Twining, "The cover-up is the disclosure and the disclosure is the cover-up." In other words, many measures of evidence indicate an *autonomous* public disclosure is occurring—an extensive process likely to be a necessary prelude to anything official ever being exclaimed about the issue.

Back in the 1950s, the public *would* probably panic if President Truman had announced he ordered "flying discs" to be shot down, or that ETs were performing cattle mutilations, or disc crash-retrieval technicians were mortally infected with some exo-biological pathogen. The psychology of social engineering would dictate the prudence we have experienced. The ufological plus *expolitical* evidence today, however, indicates a "sea change" in why no Official Disclosure has occurred. As Col. Corso is quoted saying, the issue is dominance and product development; the issue is the national security status of economic militarism; the issue is—simply—business as usual. And given an increasingly comprehensive perspective in UFO research, there is no practical purpose to a cover-up other than preventing economic disruption...unless it's to create threats that don't exist. It would be naive to perceive otherwise—given our prevailing consensus *paradigm* of conflict resolution and economic ambitions. And, we know from the above citations, *consensus* is simply a *way* of perceiving according to biased authorities, those who have decided what is acceptable to perceive.

Such conclusions are not presented in a view of pessimism. Clearly, there are many orders of reasoning as to why an ET disclosure is not officially in the offing. Our two primary exhibits for explaining why this is not plausible are the Uhouse and Leslie testimonies. However, given our comprehensive resolve for pattern recognition—our critique of *"disclosure"* would be remiss in failing to also cite *The Brookings Report.* This submission to Congress touches upon ET disclosure.

From the vantage of institutional reasoning, the *Brookings* study was "Prepared for the Committee on Long-Range Studies of the National Aeronautics and Space Administration [and the 87th Congress], by the Brookings Institution." The full title is *Proposed Studies on the Implications of Peaceful Space Activities For Human Affaires, December, 1960.* This report is probably the most readily accessible, and official, document for publicly disclosing why there has not been an Official Disclosure about ETs; i.e., it is in the Congressional Record.

Essentially, *The Brookings Report* was a guidebook of recommendations for NASA to follow. It was "intended to emphasize" necessary capabilities in managing "the potential of space activities to affect society for better or worse." NASA assumed the role of ushering society into the *"space-age."* The Congress, and/or the agency solicited *Brookings* to counsel it on preparing for how "the space effort" would impose social change throughout the world. If, as Marshall McLuhan instructs, *media* are technological extensions of man; with services and disservices—*The Brookings Report* accurately anticipated those services and disservices with which space-age media would challenge humanity. All environmentalizing innovations alter traditional realities and war-mindedness will accelerate this.

Brookings clearly indicated that a primary disservice of putting "man-in-space" was possibly discovering extraterrestrial life. Sci-fi/rationalist skeptics of ufology do not seem to have ever fully appreciated this position. Why, if ET facts were known, the public would not be overtly informed of it. In other words, the advent of officially authorizing an institution of "space activities for human affaires"—with hindsight—retrofits the history preceding NASA, with a framework for managing a possible ET-presence. NASA prepares society and itself for what Ufology already knows. The proposition made by 60 years of ufology is that hindsight shows we already discovered ET life. So, techno-culture has been playing catch-up ever since.

Today, we are decades beyond *Brookings.* With this hindsight, the reasoning expressed in the report is at least a reflection, a rear-view mirror—if not a template—for what we have seen as the de facto non-disclosure policy.

Society has been evolving rapidly through a "period of transition" to the space-age of technological transformation; or, at least acquiescing to it. *Brookings* comprehensively prefigured this process. "The space effort" of enveloping spaceship-earth with an instant environment of satellite telecommunications would bring futuristic socio-economic change to our present, although, the space-program appears to have de-evolved somewhat since the Apollo Lunar era. *Brookings* anticipated the greatest space-age impact on society as coming from two plausible sources: First, rapid technological innovation. Second, the implications of discovering extraterrestrial life.

Brookings has also stated its limitations: "The resulting interaction between space and society would be comprised of factors too broad and too complex for useful speculation in our study. So, too, with such developments as *antigravity,* a face-to-face meeting with *intelligent* extraterrestrials, or an intensive pannational attitude favoring all-out scientific research..." (as a byproduct of space research)[italics added]. The ultimate paradox is that ufological evidence shows how all three of these latter developments have already been pursued—but not in an overt fashion, disruptive of pre-existing socio-economic realities, then or now.

Many are the reasons for this de facto policy. But the purpose of *Brookings* was directed toward people mostly "out of the loop." Thus, *Brookings* posed the question, "How much research should be done to determine what research needs to be done?" In our period of transition, their "problem area" approach would guide NASA toward "effective methods to detect insipient implications of space activities."

To accommodate such social implications, *Brookings* noted that "social engineering efforts [are] required to develop reliable...personnel." And "that it seemed highly unlikely...that lunar colonies and manned flights to Mars will be more than newspaper headlines...during the next two decades." *Brookings* was correct! Now it is 46 years into our period of space-age transition, and NASA has fallen prey to a retrograde stasis: i.e., the Shuttle and a space station for what purpose? On his weekly radio program *Explorations* (July 24,2005), renowned physicist Dr. Michio Kaku questioned the purpose of the space station. Dr. Kaku observed that, with little if any science to show for it, NASA will have spent nearly $100 billion on completing the space station. But among his sources in the science community—Kaku says, "no one knows why."

This anecdote is salient, because the most plausible answer to Dr. Kaku's question may be logically found among the classified objectives for weaponizing space. *The Brookings Report* recognized the military objectives in space. *Brookings* credited a survey of business executives on their attitudes, then, toward space objectives. Ranking second, on their list of reasons for supporting the space program was "control of outer space for military and political reasons." This would be a predictive correlation; i.e., it's internally consistent with all political-economic implications cited above. Albeit that *Brookings* pays lip service to peaceful uses of space. Currently, U.S. space weaponization advocates at the U.N. have been reported to be blocking international efforts to preserve space for peaceful activities.

Ultimately, *Brookings* has intrinsic relevance to our focus on ET *disclosures.* Yet, the candor with which the report addresses a discovery of ET also applies to the pre-requisite learning curve of society acclimating to the space-age. In practical terms, the two are parts of a whole—a dissolving of traditional social boundaries, and the co-creation of a new paradigm.

On page 219 of *Brookings*, footnote three says, "What we think as individuals and as communities, and all the patterns of our behavior, make sense in a *traditional* context. They are relevant to the traditionally established circumstances in which we live. As these circumstances change, our thinking and behavior have to change too...But tradition can change only slowly and painfully...What threatens when the pace is stepped up...is moral and intellectual chaos." This is the whole issue of paradigmatic or cosmological change versus ET disclosure in a "naughtshall." Who is to say this particular "problem area" does not characterize all of the 20th century itself, and its rapidly exponential techno-social changes? Footnote six says, "we are also going to have to deal with the dangers of mass insanity...dangers that we may not recognize even in their realization...Perhaps we should...survey the whole human scene...and...preserve the heritage of humanity through the period of transition." Yet, these views simply respond to man's emergence into the technological preparatory phase of space activity—ETs notwithstanding. Ufological evidence thus indicates that the consensus control over our paradigm shift leaves it dysfunctional. It suffers from censorship by omission.

"On the matter of extraterrestrial life," *Brookings* offers the following considerations (p. 225, footnote 34): "The Fundamentalist...sects are growing apace around the world and...can pile up a very influential following in terms of numbers. For them, the discovery of other life—rather than any other space product—would be electrifying...some scattered studies need to be made both in their home centers and churches and their missions, in relation to attitudes about space activities and extraterrestrial life. In the event of its happening, of the discovery of extraterrestrial life...it is very important to take account of other major religions.

"If superintelligence [in ETs] is discovered, the results become quite unpredictable.

"It has been speculated that, of all groups, scientists and engineers might be the most devastated by the discovery of relatively superior creatures...Advanced understanding of nature [by ETs] might vitiate all our theories at the very least."

In point of closure on the ET disclosure question, *Brookings* recommends "Continuing studies," page 216: "[R]egarding the possibility and consequences of discovering intelligent extraterrestrial life...empirical studies...might help to provide programs for meeting and adjusting to the implications of such a discovery. Questions one might wish to answer...would include: *How might such information, under what circumstances, be presented to or withheld from the public for what ends? What might be the role of the discovering scientists and other decision makers regarding release of the fact of discovery?"* (Italics added.)

For those of us who have actually lobbied for ET disclosure in the halls of Congress—the questions here posed by *Brookings* symbolize an eerie ex post facto record of irony. UFO studies indicate these questions were answered in a cryptic and vexing manner, long before *Brookings.*

Ufologists are confident that the *Majestic* documents have been forensically authenticated. If so, the *Majestic* personnel are those "discovering scientists and other decision makers" to which *Brookings* refers (?). We thus know what the answer to the above questions were, prior to *Brookings* candidly advising NASA to ask them again. Perhaps, this is so NASA and the rest of us who are 'out of the loop,' can begin reconciling what actually is going on here. It is another paradox! The answers were that there would be *no official release* of ET facts. This is because the facts meant too great a quantum leap to take all at once—into a new paradigm of cosmic awareness.

To officially disclose or not to disclose is an issue of incompatible paradigms, or world-views. Those advocating an officially authorized validation of UFO research are using a mode of logic which does not comport with the common sense driving the ideological doctrines of global economy. On the other hand, an officially sanctioned ET does not comport with the rational reductionism of Left-liberals either.

Linda Moulton Howe is correct when she says we, the researchers, are "getting closer to showing the public [ourselves] that these pieces are beginning to fit together in patterns that give us insight. Both from the government's point of view, and from the physical point of view."

But when she says, "It just needs that official authorization from the government," she is actually petitioning the government to sanction a *reality* in which we coexist with ETs. And such a *reality* is one which we are still in the process of co-creating...a process which must currently include *all* the permutations of pro or con engendered by the issue!

If we did receive some sort of Official Disclosure—what can we expect to learn in such a concession? Perhaps, we can expect what we already have more than enough evidence for. That there exists an ET basis for reverse-engineered technology. But because we request access to a military or national security secret, no physical evidence or agency documentation can be expected to receive declassification, etc., etc.

There are, thus, lessons about why to build a movement of education, which does not have the immediate goal of gaining access to what security doctrines preclude. This is a lesson of integrating public evidence into a structure of perceptual evolution. Today is a period of looking at the long term, and the building of an alternative paradigm to accommodate the evidence.

This is because we are not going to win in the short term—if winning means to have first-hand access to physical evidence, which circumstantial evidence indicates does exist. This should be appreciated so the community does not get disappointed when, next year we do not find the W. Bush Administration giving security clearances to ufologists for inspecting crashed saucers. Without a more evolved worldview, we are not going to be in charge of evidence management. The same holds true for mundane issues in crisis today. Thus, we have to look 10, 20, 30 years down the road: a road of altering the paradigm, in a time when fundamentalist forces insist on preempting perceptual evolution.

Consciously or otherwise, ufology is "pushing the envelope" of social and cultural conformity. But, ultimately, ufology of the moment will be wise to also oblige itself with a progressive task. The best definition for this challenge would be institutional self-realization. As Col. Corso's book reveals it, ufology has yet to reconcile *The Strategy* that Corso affirms. Chapter Six of Corso's book is entitled *The Strategy:* a system in which the public and ufology unconsciously participate. This proposition of Corso's is crucial to our focus on the M.I.C. here, mainly because the system he alleges would be the source of the data that ufology holds to be most credible. And this is the point of origination from which *"The Strategy"* was contrived.

If Corso's *Strategy* can be cast as a credible program—then ufology plays a sociological role, for which it is not claiming to be overtly mindful. It has assumed a role—in its own becoming—by default, in a system beyond its control.

A skeptical but unbiased assessment would conclude that a system or *Strategy* of social acclimation/diversion has manifested inexorably, i.e., whether Corso is or is not correct, providing that ETs are present on earth. If you concur they are, read on...

PART TWO

An Interview with Melinda Leslie

On Covert Intelligence Involvement with ET-Contact Witnesses—2005

Presented here is a discussion with Melinda Leslie. For many years, she has investigated evidence for *Covert Intelligence Involvement* in surveillance of ET-contact witnesses. The purpose here, in articulating these circumstances, is providing insight into a twofold proposition: *One*—the logic of having discovered possible Intel-involvement in the lives of certain private citizens. *Two*—charting the parameters of public *policy* connotations posed by the greater ET/UFO issue. Point *one* can be predictably recognized as an applied function of point *two*—given the preeminent pattern availed through 59 years of ET-data collection. This pattern establishes representatives of the U.S. military-industrial structure as filling formative roles in the unfolding saga.

Background

Presently, a fair and accurate status report on UFO studies can claim an increasingly coherent body of evidence. The coherency derives from correlations occurring consistently as patterns between anecdotal, documental, and physical evidence. Culturally—like any other actual set of circumstances, the facts of an ET-presence are symbolically evident in all social arenas.

The most lucid and consistently validating documentation of an ET-presence are official F.O.I.A. releases, correlating with archival or "leaked" *Majestic Twelve Project* files. People officially representing the agencies of origin for these documents never acknowledge their existence, of course.

If the above *is* fairly accurate, then the reason for no official recognition of the history is borne out of long term political implications; i.e., the significance is an issue of global proportions and security. In an ET-presence context, "security" means re-enforcing and buttressing the consensus *paradigm of perception,* upon which, the ethos of global resources management economically depends.

The human-involvement circumstances, discussed and emphasized below, are an area of investigation conspicuously neglected by the UFO research community. The purpose of emphasis here is to show how a bit of inductive reasoning, based on all the above, would logically lead to the cases of evidence called: Covert Intelligence Surveillance and Monitoring of ET-contact witnesses.

The documental-category of evidence is most indicative of the rationale for a surveillance and control program with Intelligence involvement. And the body of compiled evidence for Covert Intel-involvement confirms both the credibility, and the logic, of *Majestic Twelve* references to intentions of reverse-engineering ET-technology. Such research and development pursuits, like any common weapons program, would only be viable under absolute secrecy.

In other words—for all people advocating the UFO/ET issue, who thus demand an Official Disclosure of the facts—the facts do not in any way bear out a basis for concluding such a *disclosure* scenario. What the facts implicitly provide is a basis for observing what has been reported: a policy/program of monitoring; surveillance of evidentiary witnesses; and *media* spin-control of the available facts. The current 59 year old policy of *non*-disclosure is a closed system, by definition. Alleged "insider" accounts of this system identify the insurance for keeping it closed: a private corporate structure has been ordained as custodian of the "issue." As retired insider Bill Uhouse termed it—"It's not the U.S. government...I call it *The Satellite Government!"* Col. Corso identified it as "nothing less than a government within the government, sustaining itself from presidential administration to presidential administration regardless of whatever political party took power." And it is in this vein that Major Ed Dames challenged Melinda Leslie with a dissuasive caveat. Following his 19 Feb. 1997 lecture for Los Angeles MUFON, Dames disclosed to Leslie "*the strategy*" she confronted: "You and people like you should drop out or you'll be hurt... I promise you, it will be a never-ending source of frustration, because you are in a closed information loop."

Thus, a contingency of "plausible deniability" obtains—if needed. And public opinion may tend to define the policy of non-disclosure as a "cover-up." Yet, the "cover-up" thesis fails to clearly apprehend the purpose of evidentiary "leaks," in the forms of documentation, and unofficial verification by former "insiders."

Inexorably, confidential facts would be expected to "leak" out. However, the *Realpolitik* of ET-nondisclosure insures that info entering the public domain systematically conforms to skeptical biases of *media* "spin." The most recent exemplar case in point was the ABC Special *UFOs—Seeing Is Believing,* hosted by Peter Jennings, 24 Feb. 2005. For 30 years network TV has acclimated public perceptions to UFO/ET details via specials, and long-term series like *Sightings* and *In Search Of.* What was censored by omission is our topic here.

In an effort to remedy such disinforming biases—Melinda, please define what you call the monitoring and re-abduction of witnesses by intelligence teams, and itemize your categories of evidence for such cases...

Melinda: To define my research is to say it is some of the greatest evidence for the reality of the alien-abduction scenario. Some abductees—not all by any means—but many report *also* having varying degrees of apparently covert-Intelligence, and military or paramilitary involvement. They have, at the very least, low levels of harassment and surveillance in their lives. This increases further to deeper levels of involvement. Cases report everything from direct low level confrontations, to higher levels of helicopter flyover harassment; phone tampering with interrupted conversations; tampering of their mail.

"Then, it rises to the level of cases actually being watched, being followed, and then being approached in direct confrontation. People show up at their home, or show up where they work, confronting them; people threatening abductees away from the subject! Telling them not to talk about it, or, asking questions regarding their experience. Quite often, these are threatening confrontations.

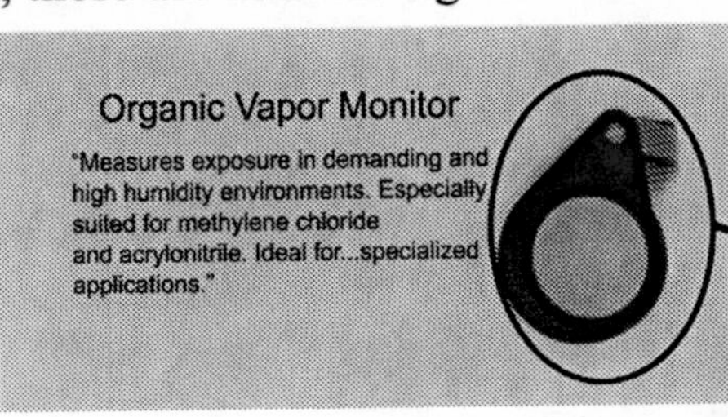

Device similar in both appearance and description to one worn in Melinda Leslie's covert-intelligence medical procedure

rendering by Bill McDonald

Evidence of break-in to home as a result of military abduction

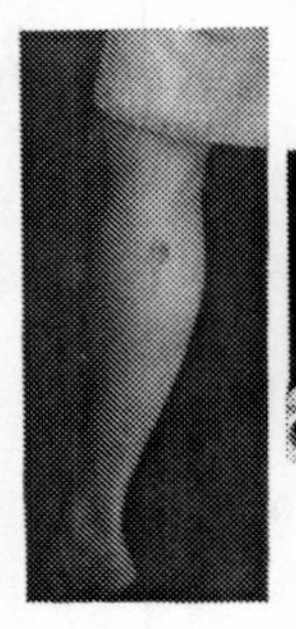

Further evidence – bruises received and surveillance technology used

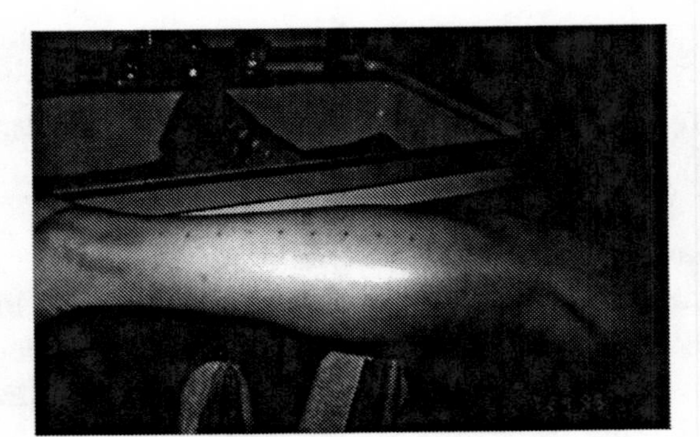

Typical physical evidence as a result of an ET abduction experience – i.e. evenly-spaced leg burn marks

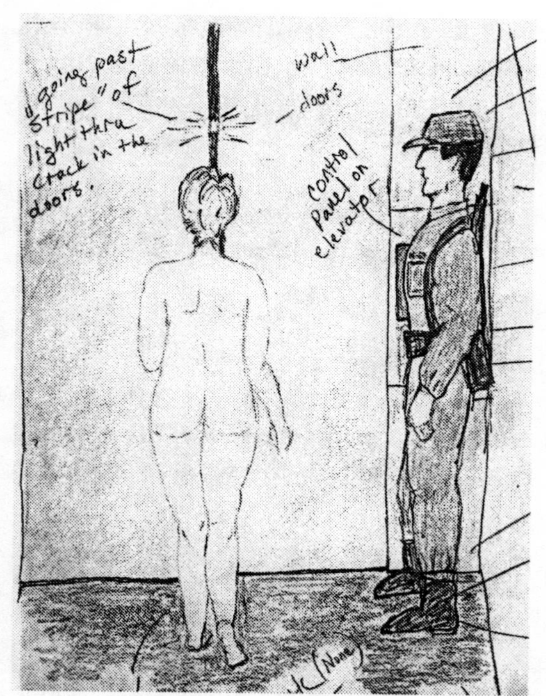

Underground base elevator ride
illustration by Melinda Leslie

"Ultimately, there is actual direct case involvement, where there are what I call *re-abductions*. Where the ET-experiencer then may get re-picked up by these either covert-Intelligence or paramilitary types, but definitely humans. The witnesses are *taken* to a human environment, and often interrogated about their experience. Often you get a threatening nature—telling them not to talk...about their experience.

"There are frequently certain questions: everything from ET motives to ET technology; both of those are hit on a lot, those subjects. Then, even a further level of involvement I'm finding, is a deepening control in their lives. Their lives start to be more manipulated. They have someone in their life; one abductee coined a 'minder': someone who is there in their life to *mind* them, to watch them, to surveil them. 'Minders' take a role in their life. This is someone who befriended them. It can begin as an interesting relationship with an 'insider' that comes up. But there seems to be other motives by the 'insider' where, on the one hand, he's providing information to the abductee; on the other hand, he's also *getting* information. Or, a combination thereof, where it's all about either he wants to retrieve info, or wanting to be overly helpful. Often, there's a controlling measure about this, too. The 'insider' influences the abductee into the direction they go with the witness' own personal investigation, or with their going public, etc.

"In addition, there's manipulation in the form of financial control. A lot of cases having deeper involvement feel their finances are manipulated so they can be controlled and limited. And there's some pretty good evidence for this happening.

R.K.: Does it seem the *"minders"* or monitors are applying various persuasion techniques to primarily prevent abductees from divulging their contact experiences?

M.L.: In some cases, yes. There's a mix of this going on. My point being, it's clear to the witnesses, the one or more monitors in their lives are involved *because* my cases are abductees, because they have those experiences. In some cases it's encouraging to the witness; in some cases it's very discouraging.

R.K.: OK, then, overall the *system* is heterogeneous. I.e., it's not designed to apply one limited approach or tactic of persuasion.

M.L.: Usually, it tends to be more discouraging. In fact, I'll quote one witness who quoted her *"minder."* The person involved with her claimed they were involved in the UFO subject and the UFO community "to derail the train," quoting the *"minder"*...But, I also have some cases reporting people involved who seem so forthcoming with information, and so encouraging—you can't just say there's one agenda.

"There also seems to be deeper levels of control, *influencing* what witnesses may say about it, the way it's said and the kind of info given. Like, they're being influenced by encouraging them to talk about certain subjects; while maybe not talk about others.

"Again, my point is there's this deeper level of involvement. In some cases it can be where the witness starts dating someone with Intel-involvement. Or, just having someone who's highly placed in the military-industrial complex (M.I.C.), or in military intelligence, or ex-CIA, or NASA, or other agencies that suddenly show an interest and become friends with them. Someone suddenly cares to become their friend. When, prior to having their contact experiences, this monitor-type wouldn't normally be someone the witness *could be* friends with. And certainly these monitors seem to show a sudden, above average interest; a little too supportive sometimes. It's curious.

"Even in *my own experience* I admit to having this. But I've come across it in other cases. And there's this dating with individuals involved. At times, the witness reluctantly becomes associated, not just romantically, but simply in a friendship. Where the witness may have one of these people who constantly attends events they have; constantly calling them on the phone. And this interested party may be someone who the abductee is suspicious about."

R.K.: OK, in other words, a pattern occurs with the ultimate background of these *"minders."* Whereas, ostensibly they begin by showing up in some plausible pretext. The commonality between cases is an introductory, mundane interest in a witness' life, leading to the witness divulging their abductee experience. Then, there's a realization that this new acquaintance shows a very unusual, if not dedicated, inquiry in their "experience." Ultimately, this reveals the pattern where these inquisitors are "insider"-type people, of varying degrees: they have a background lending itself to the relative scenario and purpose of the agencies they actually represent. I.e., the Military-Industrial Complex, et al.

M.L.: Sure, sure! Absolutely. And the witnesses, wanting to find someone *to confide in,* to talk to—they want someone who understands, is supportive, and is forthcoming with information, which maybe provides details and answers. Helping the witnesses understand what's happening to them. Just having the feeling you've got someone providing you [with] information, and verification of your experience, leads you on to be very forthcoming with personal background. You know, the question then becomes: who are these *"minders"* working for; where is this info going? I don't clearly know the answer to that.

"But all these cases indicate motives going beyond just the harassment and surveillance level, which is their initial objective.

Healthy Skepticism and Discernment

"At square one, you have people who feel or believe they've had abduction experience. They usually conclude this based on the memories, or recalled memories; also, a certain amount of anatomical evidence. Often they feel this has occurred periodically throughout their life; i.e., more than once. Again—this may not be a person who even goes public. Yes, such attention would seem obvious for an abductee who's gone public. But, in many cases, these are people who are very quiet about their experiences. They have not gone public!

"I've been dealing with a new case just referred to me. She has not gone public; not known, doesn't even know much about the abduction subject. Yet, is having this happen! And she says, 'This is gonna sound crazy, but I'm having...', and she'll tell me something. Of course she's telling me something I've heard a hundred others tell me, because I've worked with over a hundred people with these experiences; probably over 40 people quite in depth.

original illustration by Melinda Leslie

rendering by Bill McDonald
Melinda Leslie's Medical Procedure in a Haz-mat Type Tented Environment

R.K.: Regarding the thoroughly anonymous witnesses: why would they have invented this scenario? At this point in time, after a number of years gathering the anecdotal cases, we find there is a very important pattern here—in the sense that experiencers have attracted the attention of the "monitors/insiders" or whoever they are. Albeit that *re-abductees* are not visible on "the radar screen," as they say. So—how were they detected and identified as being A: abductees; and B: therefore qualifying for this unusual attention they're receiving?

"Then, ultimately you've been introduced to quite a few of such cases. The point I'm emphasizing here, are unusual patterns. And this can bring us back to the intro here regarding *media.* Because the importance of these type cases—quite to the contrary of media exposure, etc.—in *media,* these type cases have not been emphasized; abductions, yes. Not *re*-abductions by human Intel.

"Maybe even more important—in the work and publications amongst the prominent abductions investigators—they have gone out of their way to either play down or intentionally choose not to emphasize re-abduction cases."

M.L.: Yes, that's right.

R.K.: So, if we were to speculate as skeptics would—let's say these mostly anonymous witnesses have decided to pose as victims of two things: both abduction, and also, being harassed and kidnapped by humans, about their abduction—even though, there's not a lot of attention publicizing re-abduction/harassment, generally. Thus, the abductees would have to be seeking your attention, who is investigating the scenario. And this is actually even more unusual than the skeptically-dubious ET-abduction scenario. In other words, if re-abductions were fictitious, what is the seemingly unachieved goal? I.e., there's no payoff. What would the payoff be?

"*Point One:* The witnesses are not trying to publicize themselves. *Point Two:* You say they are obviously embarrassed about the weirdness of the monitoring in any case. And they are in fear of being classified as paranoid, etc. So, the area of publicity is not doing too well since there aren't that many people to tell. And you've discerned that the prominent or professional investigators are not too interested. So, the witnesses seem to find you, Melinda, as a last resort."

M.L.: And it's not so much they want to simply tell someone. *They want to understand.* They find out this has happened to others, or that there's someone researching this overall aspect. And most of the abductees who have this happen—at any of the levels of involvement—when they have any of this happen, they're pretty much certain they are the *only* one.

"Then, they wonder, 'What is it about me; or my case; or my experience that makes me of interest to these guys?' They begin to question themselves, question their sanity. They have this dichotomy: 'OK, well this has happened. I've got these guys involved. So, is this evidence for the fact I've had something very real happen?' And there's the other part saying, 'Well, these guys can't really be causing all this to be happening.' And they start thinking they're paranoid.

"So, when the cases read an article, or are referred to me, etc., they go, 'Oh, my goodness! I'm not alone!' So, usually they want to talk to me just to find validation and verification; to understand they are not alone, that they are not crazy...and it's happening to quite a few.

"All the major researchers have at least some cases of Mil-Intel involvement. Some of them have multiple cases.

Deducing the Obvious

"I've so far worked with cases of ET-abduction who are quite average in every other aspect of their lives. They have nothing going on to warrant monitoring or an interest by the government, or special-ops, or covert-Intelligence agents—*at all!* And, when it comes down to the *way* they're being monitored—they are surveiled and harassed, and followed *directly after an experience;* and directly in relationship to if they've gone public, *or if they are thinking about going public.*

"Of course, the questioning, the interrogation, or the *'minding'* and befriending someone is in *direct relationship* to questions and information about ET-experiences. In other words, there's nothing there to suggest any other reason for the interest, other than their abduction.

"Some remaining categories of evidence for the Intel-involvement are: agents taking photos; eyewitnesses to re-abductions and harassment/surveillance; break-ins with photos or documents missing; bruises after an experience; other physical symptoms; photos of helicopter fly-bys; being drugged with medical lab tests; seeking an installation location you were taken to, and later verifying its existence."

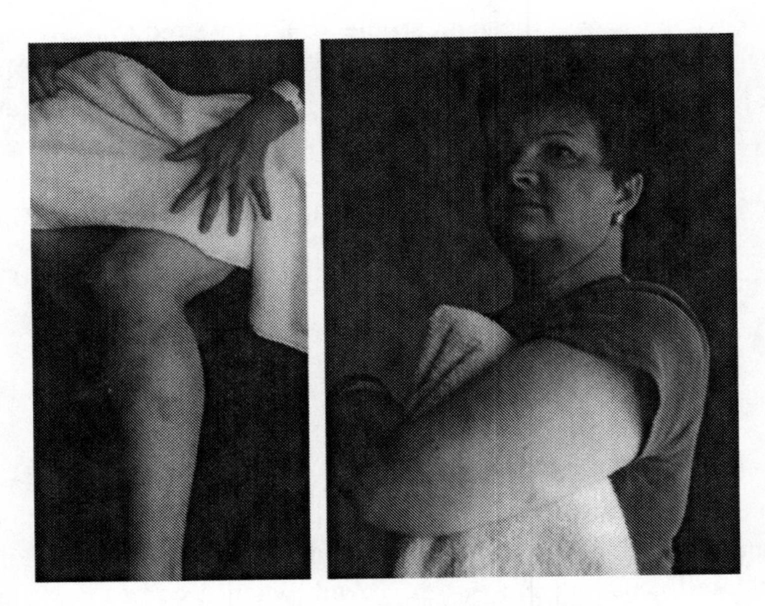

Melinda's bruises after being abducted by covert-intelligence agents - 2001

R.K.: Yes—due to the obviously systematic methods you report, a clearly military-industrial set of motives are at work in your cases.

"Deductively, this is likely. Since, the most persuasive Lowest Common Denominator in all ufological study converges on various military correlations. The most reputable was the Col. Philip J. Corso account of crashed ET saucers. Likewise, it's perhaps predictive, or *a priori,* that if an ET/crashed saucer/*MJ 12* nexus is accurate, a scenario such as your own would surface.

"Clearly, all your facts are internally consistent with the uniformity of data amassed over decades: that the *sub rosa* military-industrial milieu holds the best ET evidence."

M.L.: Going back to the issue of technology—again, based upon crash retrievals, based on the military wanting to build craft, etc., I still don't want to paint this with too wide a brush—but there does seem to be a military mindset at work. In that, the group running this covert operation or task force is military in nature, is paramilitary. It's military bases; it's known underground military bases; it's science installations that have military purposes; it's guys in uniforms and military equipment. I mean, it's everything, about what the abductees are reporting.

UNIFORM PATCH SEEN ON THE SLEEVE OF A MEMBER OF THE MILITARY WHO ABDUCTS HUMANS WHO HAVE HAD ALIEN ENCOUNTERS. THE MILITARY ABDUCTS CITIZENS TO RETRIEVE ALIEN IMPLANTS AND TO LEARN ABOUT THE ALIENS, BY INTERROGATING THE CITIZENS THEY HAVE KIDKNAPPED.

Possible military uniform insignia as drawn by an abductee

"Then, when you go back to the study of crash retrieval cases, it seems to be consistent: crash retrievals were handled by the military. The ET technology went to the military for military purposes."

R.K.: Well, it's the military-industrial interlock, you know, with its "revolving door." And the industrial actually does the development.

M.L.: Yes. So based on that, if there's a military pattern of interest—and I return to the abductees—what is it they are questioned about? *They're being interrogated about how the ET technology functions* and especially some of the more exotic functions.

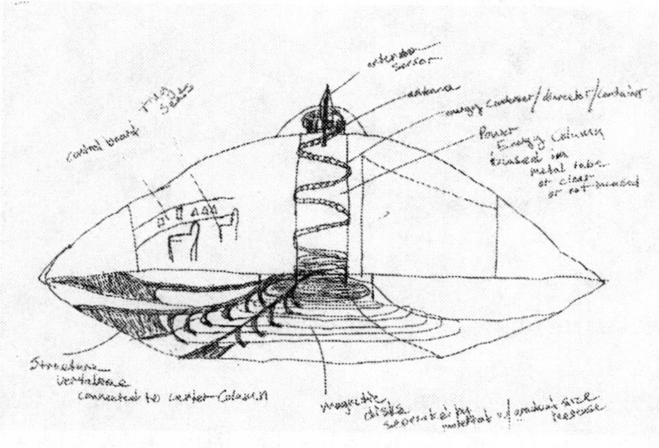

Melinda's original drawing of ET craft propulsion system

The Mental Interface Objective of 'Reverse Engineering the Abductees'

"In their development of these more exotic alien-derived technologies from crash retrievals, or from various ways they may be learning of this, if they're receiving info other than from crash retrievals alone—but regardless, there seems to be this aspect of the direct-interface between the alien or person, mentally. In order to be able to navigate, operate or fly alien craft, you must employ this direct-interface capability.

"And the abductees realized, too, that in the ET technology there's a certain genetic component to be able to operate the technology. This has everything to do with the ETs. [i.e.], in some of the technology you have to be an alien in order to operate it.

R.K.: Yeah, I recall you explaining how an "insider" revealed to you that answers about the design of ET craft were derived from alien autopsies; e.g., in the Bluebook 13 report.

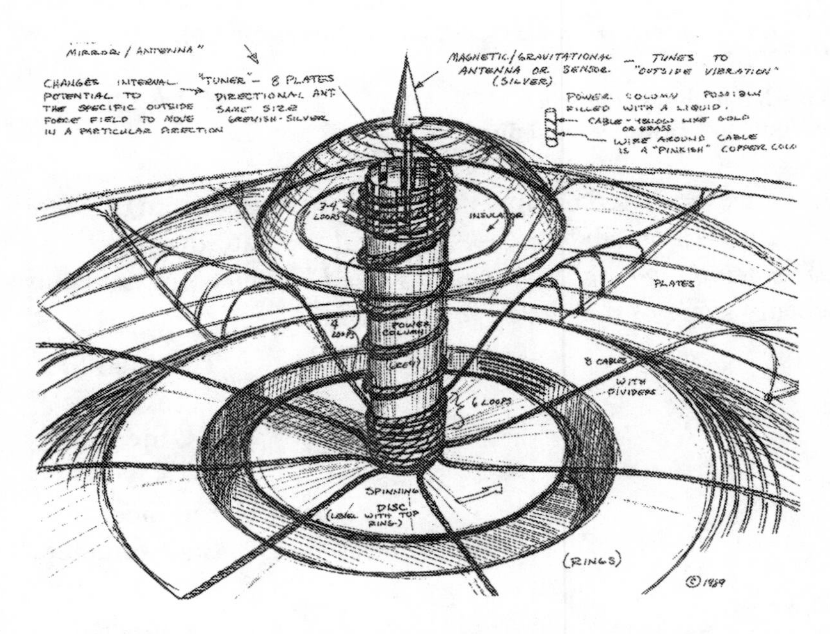

Technical rendering based upon illustration above

M.L.: And they realize there's a physical interface requiring a certain amount of alien genetics. There are obvious things: from size, and the way their eyes function, or how their hearing may function. Even down to ways that they may function mentally. And in M.I.C. development of this, I think they've come across some blocks: incompatible areas they're trying to get through in this understanding.

"Now, one thing—going back to basic ET-abduction experience—is telepathy. This is direct-*mental interface*. The ability by the alien to kinda give abductees, in one whole thought—if you will—how to operate a piece of technology. Then the person perceives they can operate the technology.

"There seems to be a lot of this in abductee experiences. Reports indicate a lot of this paranormal aspect to the functioning of the technology, to the ability to interface with it. The technology seems able to read your thoughts! Just as the ETs reportedly communicate that way, there's a direct communication with the technology itself, i.e., in that same way."

R.K.: So—another correlation. Collectively, the evidence suggests the experiencers routinely perceive they were, by definition, partaking in a "learning curve"! Wherein their period of adjustment in having encounters, they progressively learn parapsychological behaviors.

"Coincidentally, in the area of your human interference cases, this is distinctive: one of the primary objectives of the monitors or interrogators is info about this very same psychic interface necessity, between user and techni. Of course, IS this a fortuitous coincidence? One between witness reports of their inquisitors being really interested in their psychical knowledge—and—tech-journal projections toward human-interface R&D goals? Perhaps it's no coincidence at all."

M.L.: Exactly! And you can extrapolate even more from that—it's also the psychic abilities of the experiencers themselves: everything from remote viewing to other various psychical abilities. In their claim to produce these advanced mental states, there's an interest in the abductees in that way. The interest is how they interface with ET technology; how did they react to it, what were their reaction times? How did they know what to do? Was the technology itself instructing them, or ETs instructing them?

"Also, returning to possible genetic components to this—ET abilities infer a genetic determinant for what seems natural for them. And there is [or seems to be] something about abductees that may make them predisposed, genetically, to be able to do psychical procedures.

"Now, the truth is, I don't know if that is a chicken or egg question: if it's something about abductees, generally to begin with, making ETs *contact* them? Or, if the ETs have done something to us genetically to enhance our abilities? Probably, it's a combination of both.

"Is it that ETs are randomly selecting people—no! There seem to be some common traits among abductees. Researchers generally agree, there is a pattern of personality traits, and possible genetic makeup to abductees; consistently, this being part of the ET scenario."

R.K.: You mean the primary abduction investigators have identified a pattern of their cases following family lineages?

M.L.: Yes! Family histories. This suggests there's been a continual monitoring of a genetic heritage.

R.K.: This family pattern would not be easily explained if abductions were totally fictitious. This is also what led therapists to conclude, whatever "abduction" is, it is not random, right?

M.L.: Exactly. So, figuring the time displaced throughout a family lineage—has there been a continual genetic manipulation, in order to achieve an ultimate end?

"All the above is evident in basic abduction research. And certainly these Intel-forces, who are spending so much effort on tech R&D, plus genetic research—logically, they would ask 'how can we benefit from it?'"

R.K.: There is black-ops genetic research vs. *mainstream* commercial genetic research; there's commercial tech R&D vs. black-budget R&D. Pulitzer Prize winning reporter Tim Weiner wrote his book *Blank Check* on black-budget operations. But the principle of connectedness here is the element of time. Where the commitments to exotic phenomena are congruent among those respective facts, defining the 3 groups. And by each having such a thing in common, they are congruent with each other over a long period of time.

M.L.: And there's also now evidence suggested in my research, this covert Mil-Intel involvement with abductees is not new. It's been going on over a period of time. So, as you can say the ET-abduction has been happening, and well documented for decades, you could say for the same amount of time there's been the covert-monitoring aspect. It's logical! As intelligence forces learned more, they would want to know even more, as their tech-methods evolved in capability to pursue more.

The ET Presence is an Exo-political Influence

R.K.: If we do have evidence for a clearly concerted effort, by military-industrial Intel, to pursue info detailed here—the illicit methods of unsolicited interrogation or monitoring must impinge on the UFO Disclosure policy. The UFO community has found, historically, an unofficially gestured policy of Non-disclosure; i.e., a no acknowledgement of the ET/UFO issue at all, policy—period. Yet, the best evidence for an authoritative awareness of the basic ET issue is: voluminous pages of *official* F.O.I.A. files. The UFO community takes pride in publicizing this evidence.

"Necessarily, there's a close synergy between the industrial side of the Intel equation, and, the military side of the Intel equation. But, why do officials in authority absolutely refuse to disclose the facts *they* declassified? The issue does not exist for officials.

"So, for all the people in the UFO community who wish to organize the community politically, for persuading an official disclosure, to no avail—Melinda, what do you think your information holds, for providing an insight into the historic policy of refusing to officially recognize any of this ET issue whatsoever?"

M.L.: I think, if you reduce that to a single most-common-denominator, taking a step back to see it better—it's because there is too much to lose by disclosing it; there's too much *to gain* by continually keeping it secret.

"Obviously, there's a huge amount of black-budget corporate spending and justification which self perpetuates a cover-up; so much at stake regarding both loss of funding, and financial gain from the development.

"Obviously, there's a structural dynamic implemented, including massive corporate monies, to fund this. Based upon the history, it appears to be working: you know, everything from black-project technologies to development of the exotic craft people are 'sighting.' Some of which may be military-Intel pilots flying it. Certainly with the black triangles, that may be the case.

"If what we're seeing with the involvement is going on, there's a whole infrastructure of exo-political influence put in place for the purpose of implementing *the system.* All this would be based on our evidence.

"BUT—if they feel that disclosure is a threat, they may feel this because they don't have a handle on it. Not in such a way that, if presented to society, it won't cause a collapse of civilization as we know it."

Shifting the Perceptual Paradigm

R.K.: You mean collapse of *the paradigm*—

M.L.: Exactly, exactly! They don't want to collapse the paradigm. The paradigm, as a social structure, is—well, it's not really working, but it works on a level. It works in its denial of the fact that it's not working; i.e., the paradigm structure of society is the world economy and how it functions as a *set of perceptions.* There's no way to come out with this information in a way to not collapse that...I don't necessarily agree with such a policy, but power structures may.

R.K.: Given all we've submitted regarding the persuasive effects upon human behavior and technological evolution, due to ET contact evidence—if this is basically accurate, then you can't be wrong. *Especially* regarding *degrees of* benign *social affectations due to contact.* A different paradigm or system of perceptual consensus would be needed to digest ET facts. I.e., the *de facto* shift in reality has, historically, already taken place 60 years ago—but only for the handful of humans *"cleared"* to have *"a need to know!"* So, yes, that would be unprofitable to now admit to, to disclose. There's no precedent for it either.

M.L.: Yes, the *reality* of it, and what the ET technology brings to us. This is in fact where we need to go as a human family. Yet, when you change that paradigm, when it does change, things happen—it's called *revolution!*

"The power structure is benefiting from this financially. If you go back to that *corporate* power structure, you return to a system pursuing a path. Disruption of that path would mean the viability of it being too drastically changed. So, in a desire to keep the status quo going momentarily, it's a bad habit."

Approach/Avoidance Regarding Psychiculture

"Now I can get into something here. In this interface with ETs *comes a new paradigm;* a way of being, of a way of consciousness. A way where people operate as more than their physical bodies. Where, for instance, you're able to remote view, to manipulate matter with your mind.

"And I believe as we move toward more interaction with alien cultures, this paranormal way of being will increase. I believe it's already there. And it's one of the reasons the Intel-monitors are interested in abductees; *to suppress* these behaviors. Not necessarily to suppress it in abductees themselves, but because the Intel-types want to develop the psychical. At the same time, the monitors do not want abductees *to empower* other people to be able to do so. Such empowerment lessens manipulation and control: social engineering.

"So, in the interest of the obsolete paradigm, their desire not to *shift reality* results in a desire to limit people in that way. To limit the consciousness expansion of people."

R.K.: Right. The 9/11 reactionary advertisement of terrorist fear-mongering keeps collective consciousness in the 2nd chakra survival mode. Capitalizing on the "fear premium"; instilling perceptual, and economic, fascism.

M.L.: I think that's one of the driving factors. Economics alone would be reason to keep a lid on this. But also it's about their ability to remain in power, by keeping others disempowered. And I think they've developed it so far out in the black R&D world—they know when that Genie is out of the bottle, the gig is up! The world will shift overnight.

The Spin Control of Unofficial Disclosure

"Now on that same note there's some hope in the following: it also appears, as indicated in the Intel-involvement cases, that some of the 'minders' seem to also be allowing certain information out. There may be a desire by *some* on the inside to slowly bring that [unofficial] change around?! But—*it's about the control of such a release;* it's about a controlled change. I.e., this approach does seem to be present.

Yet, all-in-all, this controlled release of info is *not about an end to the cover-up!* It's just about manipulating the cover-up in a different way!"

R.K.: Well, the system itself is a system of non-disclosure. The National Security State per sé is the politicization of an occult sensibility. There's always people wanting to leak things out, whether it's *MJ 12* files, or more prosaic info.

M.L.: But that doesn't mean the paradigm creating the non-disclosure is changing.

R.K.: Our advantage here is that the information-age-environment has brought a situation where a paradigm shift is inexorable and unofficial! Yet, the *realpolitik of Non-Disclosure* is such that you can keep collecting more and more data, but it won't change the doctrine of Non-Disclosure....

Meanwhile, our issue here is why your covert Intelligence involvement cases are now newsworthy. So, let's be scientific about it. Science must be predictive.

If we take the best documentary UFO evidence, we have: Bill Moore's *MJ 12* documents, and Tim Cooper's *MJ 12* docs, and Bob Wood's *MJ 12* docs, and John Greenwald's F.O.I.A. docs, and Clifford Stone's F.O.I.A. docs, and Bruce Maccabee's FBI docs, and Stanton Friedman's archival docs, and Richard Dolan's *National Security State* history—and if we use these as a predictive basis, it would be quite predictable that your case histories would become Exhibit A. Given the authorial intent of the National Security mindset, creating these voluminous files above—our powers-that-be would be in *dereliction of duty* if they did not monitor your witnesses! Simple.

Thus, Melinda is reporting on the pedestrian, individual or personal interface with those people keeping tabs on the issue. The gross body of UFO data overall can be viewed as a basis for recognizing that

Covert-Intel Involvement is a logical procedure to be expected. In this setting, the *"minding"*/re-abduction would not be an anomaly.

A wise use of our energies and resources would be to focus on UFO facts as an incentive for consciousness raising. Because you already live in a new paradigm if you appreciate the validity of UFO/ET facts.

Melinda quotes professorial Stanton Friedman in her conclusions in relation to keeping UFO matters in perspective: "Disclosure of all the facts on the UFO subject would be the end of nationalism world-wide...that isn't going to happen, because that's the only game in town."

PART THREE

"Ace In The Hole": How Our Military-Industrial Interlock G.E.T.s Technological Superiority

The experience recounted below is a composite verbatim transcription of select public appearances by Mr. Bill Uhouse [Anglicization]. For 38 years, Mr. Uhouse declares having played a formative role in the ultimate Unacknowledged Special Access Project (USAP): the literally subterranean program for perfecting functional flying discs. If you see one, you may call it a UFO. Yet, according to Mr. Uhouse, humans have manufactured many disc or triangular-reshaped disc-craft: IFOs under our intelligent control.

Uhouse also insists that these most historic achievements of human ingenuity did not originate from exclusively human ingenuity! Empirically, such an ingression of tech-novelty could only be at the behest of an *alien brain trust.* Col. Philip J. Corso testified to this in his book, *The Day After Roswell*. Uhouse is the first hands-on, career term witness to verify Corso in detail! Following 10 years of WWII and Korean War service, as a Marine Corps fighter pilot, Uhouse was discharged as Captain. He then embarked upon a resident career at the veiled frontier of an exclusively new technological paradigm! It is a story which entails the Tao of physics: gravity controlled flight.

For students of ufology, the witness of Bill Uhouse is without peer! He rarely spoke publicly. He appeared most often at an event hosted by his friends Pat and Joe Travis of Rachel, Nevada. Clearly, the medium of print does injustice to the authenticity of Uhouse's delivery. The nuance of experience in his speech, however, evokes a sense of innate familiarity, only found rooted in experience. Here, then, is the last of those rugged individualists covertly charting a parallel frontier beyond our concurrent entrée into the obsolete Jet Age.

A Threatening Coincidence

Bill Uhouse: A couple years back, after that last *Friendship Campout* meeting at which...I attended and also spoke—what really happened was, about two weeks after the meeting I was doing work around my yard. I've got a very large place in Las Vegas. And I was cleaning up after [some construction] I finished...over the last several months. While I was picking up small timbers...throwin' this stuff in the wheelbarrow, so I could then take it to the dump, while I was doing this, being bent over, I was hit in my left arm! And after about 3–4 minutes I picked myself up off the ground. Of course, my arm, blood was running down my arm. I looked over and my muscle was hangin' out of my left arm. [Looking at it] from the back, from where I was hit...it would be about, maybe 4–5 inches from a vital organ which would be my heart.

Bill Uhouse, Rachel, Nevada 1998

"Of course, I was in a daze, slightly, and I just didn't understand what the hell happened; what hit me? I picked myself up, went down to the house, I was the only person at my home at that time, my wife was in the hospital for an extended stay. I got a bottle of peroxide out and opened it on my arm. I grabbed a tourniquet and wrapped that around my upper arm here, so I could stop the bleeding. I then took peroxide and dumped it over, as much as I could, over the back of my arm; in the shower of course, and blood was running all over!

"The only one that was around that day was a neighbor friend; a lady. I went over to her house, and I told her what happened, and brought my first aid kit, of course. And she said, well she said, 'I better take you to the doctor.' I said 'No! Let's just bandage me up and let's see what happens.' So she did a pretty fine job of it. I actually asked her to push the muscle in as far as she could, and stretch tape over it, which she did.

"I eventually did go to the doctor. But, it was about a month later—after I thought I had a little infection.

"But, after a few days of getting myself back together again, using one arm the best I could, I went back out to try and find out what hit me. Of course, there's a street in front of my house, and my back was to the street. And the projectile came down at an angle, apparently, because, I was bent over that far. Unless the guy was laying down somewhere, you know and shot at me, or...

"At first I thought it was possibly a turbine blade from an airplane, that might have come off and just hit me. Whatever it was, it didn't go into my arm totally. It went in at an angle and came back out. The wound is, I'd say about 2 1/2 to 3 inches. So, it went in one side, dug a big hole and came out the other side.

"The girl who taped me up—her husband came over. We looked around for something, to see what it was...Later on, I didn't really know until I went to the doctor, and he said 'You were shot!' I said, 'what was I shot with?' He said he didn't know, but 'You were shot; that's the indication that you've got, from the results of the scar.' He had to pull away some of the scar, to clean it up...and rebandage it. He said, 'Did you report it to the police?' I said 'no, I didn't hear anything, and I didn't see anybody, you know? I just don't know what it was, I can't prove it was a bullet.' He says, 'it indicates it *was* a bullet.' I said 'OK, but I'm not goin' to the police; you can write down on your report whatever you want.'

"After 10 days or so, I was visited by two individuals. Nice young gentlemen about in their 30s. They came in the yard. I was in the back working the best I could; still trying to clean up. They said, 'Hey, is this EBE?' I said, 'if you want it to be EBE it is.' He said, 'No, no, is this EBE, do you sell the Alien Deck [playing cards]?' I said, 'yeah, I do. I'm the designer of the Alien Deck, you wanna buy some?' And they said, 'Let's talk awhile.' I said, 'OK, come on back down to my garage, I got an office.' And we sat down; they showed me their identification. They were two NSA guys; here in Vegas.

"They didn't tell me what their business was, but asked me a lot of questions on what I'd been doing since I'd been retired...They seemed to know a little bit about me, and we had just a general conversation, then those two guys left.

"Three or four days later, a colonel—I don't know where he was from, if he was from Nellis [AFB]? He also spent 3–4 hours with me. Of course, most of these guys asked me what happened to my arm. And I thought to myself, well, if you guys are NSA maybe you really know what happened to my arm. Like I said, I didn't hear a shot.

"Someone would have to be about 300–350 yards away, otherwise there would be nothing to hide behind [other than] some bush...off into the desert. So, the great possibility is, according to the doctor, he says I was shot.

"I just wanted to kinda express that first. Because what the hell did we say up at the Friendship Campout that maybe could have got me into this trouble? There were a few things that I said, but I'd like to get to those as I go along...whoever did it, we'll probably never know. I think they were aiming for something else.

"The NSA guys also asked. I said I was hit with something. I didn't tell them I had gone to the doctor; anyways, it was sort of a coincidence!

"It's not that I'm not going to give some additional information here today, I am! I'm going to talk about the [flying disc] Simulator in regards to my presence on *the inside,* when it [the Simulator] was operating. And, some of the *phenomena* that occurs when you're inside one of those things, when it's in operation; as a simulator, not as something that flies off the ground.

"Of course, the disc-simulator is a whole lot different than an aircraft [flight/training] simulator in a lot of respects; for one thing it's round.

Reviewing the "Camouflage Through Limited Disclosure"

"I'd like to bring up something else, too, before I go on to that: I understand that Bob Lazar [and Gene Huff]...are planning a movie called *S–4,* which is supposed to start in production in August [1998]; which should be an interesting movie—moreso than some of these other movies they're putting out. For example, this *ID4 [Independence Day].* I [brought] up some things on the net...their creature that they have, or what they depict as aliens. But they're horrible looking four legged, spider types. After dissecting sort of the whole program they've put on, I felt that what they're doing now is trying to sell toys to kids after it comes out in July. They'll be selling these miniature *aliens*...involved in the *ID4* thing. The other movie they're going to put out is *The Arrival.*

"Now, something came on the net recently, after *The Arrival* was announced, it's called *EON-4.* If you look at that program *EON-4,* there are a lot of things that don't make sense. They talk about *Klaatu,* [as in] *The Day The Earth Stood Still.* They talk about these probes, identified as *Klaatu.* They show pictures of these probes [etc.]...It's quite interesting and it's very well done.

"My opinion on it is that it's more advertisement...Plus, the Bob Lazar *S–4* movie should be pretty interesting as far as the UFO phenomenon is concerned.

"One of the main interesting things about all of this stuff—in looking at some of these TV programs—and I haven't seen a lot of TV, or attended a lot of movies in my life, because I spent a lot of time doin' other things, but since I'm retired...I look at them, and I'm just wondering. *Uhh, just when are they gonna really put the real stuff in the show?!* I mean, if they got, for example, the grays; and the grays are depicted now, you know, in various literature and books [etc.]. No one has really come close to even putting those guys on; whether it's on a movie set or anything else!

"I saw this movie *Fire In The Sky,* and I heard the two gentlemen from Arizona talking about it [Travis Walton and Mike Rogers]—and he [Walton] said 'yes'...besides some aliens, he had met some human types, too.

"And I really believe—and, of course I was under the impression [due to his 'agreement'], when I had started speaking, that we were taking the cap off some of the things. I worked for this 'group' for 38 years."

"I want everybody to understand that It's Not The U.S. Government!

"You'll never get anything from Washington, D.C. Or, say you pick up some document regarding [this]. You might get *something,* but you're not going to get what you really want! Just understand what's been going on!"

**Editor's note:* in his witness testimony, submitted by Uhouse for documentation in Dr. Steven Greer's *Disclosure* publication (2001), Uhouse anticipated: "By 2003, most of this stuff will be out...maybe not the way everybody expects it. The reason why I said that is because the [oath of secrecy] document I signed ends in 2003." Quite to the contrary, *official non-disclosure prevails!* This policy reinforces punitive measures taken, some years prior, to interdict a personally authored/illustrated account Uhouse intended for the Robert Bigelow N.I.D.S. foundation. Uhouse always claimed his *unofficial* disclosures were due to an agreement he understood to have with his supervisory authorities. Higher authorities apparently objected, and changed the rules he felt he understood. Thus, the introductory story here of his "coincidental" injury.

Recruiting The Chosen

"I started with the program way back in the '50s. I was just a young guy out of the service. When I got out of the service, I accepted a job at Wright-Patterson AFB. At that time they were flying F-89s and B-47s, and [others] they were building at that time. That was the early stages of jets. When I was approached, in addition to a couple other people I was associated with, we started on a particular project. Although, it wasn't totally identified. It was not identified as, 'you guys are gonna work on flying saucers' or flying discs. It didn't happen that way. As a matter of fact, it took about 4 1/2 years for us to be cleared, after various different projects I was on, by these particular people ['the Group']. When I worked for the *Satellite Government,* it was basically out of New Mexico, at Los Alamos and White Sands, N.M., not White Sands, Utah. My portion was the redesign. For example, on the disc simulator or some other assignment, I would do the redesign work, the drawings and so forth, back at Los Alamos and White Sands. And I'd spend my time out [at S–4], 2 or 3 weeks, installing the item, and of course test it.

"I started on the F-89, B-47, F-102 simulators; ultimately the aircraft, but basically the simulators. I was hired in the early 50s, out at Wright-Patternson AFB, as a test pilot. In those days, you were required to have a BS in mechanical or electrical engineering...you couldn't be a [pilot without it]...where, back in the old days we used to have Buck Sergeants that flew in the Marine Corps. And Navy Chiefs used to fly. They had only a high school education. But after that you had to be an engineer, because of what they required, since they didn't have simulators in the old days. When the pilots came down, they needed explanations on how the plane reacted, and the pilot needed to understand the basics of an aircraft so he could properly make out a report.

"So, going back to the simulator, I was hired at Wright-Patt AFB by a Col. Reiche. I'd never met the guy during the two weeks preparing to enter the program.

"I want to bring up one point here—I'm talking about a different government—a *Satellite Government.* It's not the same as our Shadow Government, or the U.S. Government—it's a government in its own right!

"And, I'm going to say something here about people, the types of people they [the *Satellite Government*] hired. If there are some liberal whackos in here, some ACLU-types, you know, you got my apology! But they wouldn't hire any Jews, no Chinese, no Japanese, no Blacks. If you weren't of European descent, and you couldn't trace back at least 100 years into Europe—you would never get into a particular program. As a matter of fact, they wouldn't take you...and also we had some CIA people who were involved in the *Satellite Government*...and yes they wore suits—but getting back to where I was after the service: I was approached by this Col. Reiche, who invited me into his office. He was dressed in full uniform. At that time it was the Air Force; no longer the U.S. Army Air Corps, it was the Air Force. And he asked me a bunch of questions; asked me how I'm interested in a particular program that is unique, and I said, 'Sure!. Why not.' So, I thought about it a while and I decided, well, instead of staying in service, I might take a chance and go on with this particular program. There were 5 of us that got hired at the same time.

"My first assignment was at Link Aviation in Binghamton, New York, where they were building simulators. What they were training us on was to learn how to design and build a simulator. However, it was still 4 1/2 years before I got a clearance. Even though the disc we would work on, the flying disc simulator...these never leave the ground.

Simulation and Solicitation

"We initiated the program at Los Alamos; we later, during the construction period or building them, we moved it to White Sands.

"My particular design on this aspect is the flight deck. Although the all-over design of the transition or redesigned ship was provided for us to perform, it was furnished by others. They would send drawings over to us, and we would complete the design; the parts that were required for use in this particular thing were meant to look the same as the actual ship. Sometimes they'd...see what our thoughts were, what our design was before they finalized what went into the actual craft. One engineering designer's offering will be a little different than the others. What they try to do is put it together—the main thing is, it has to work!

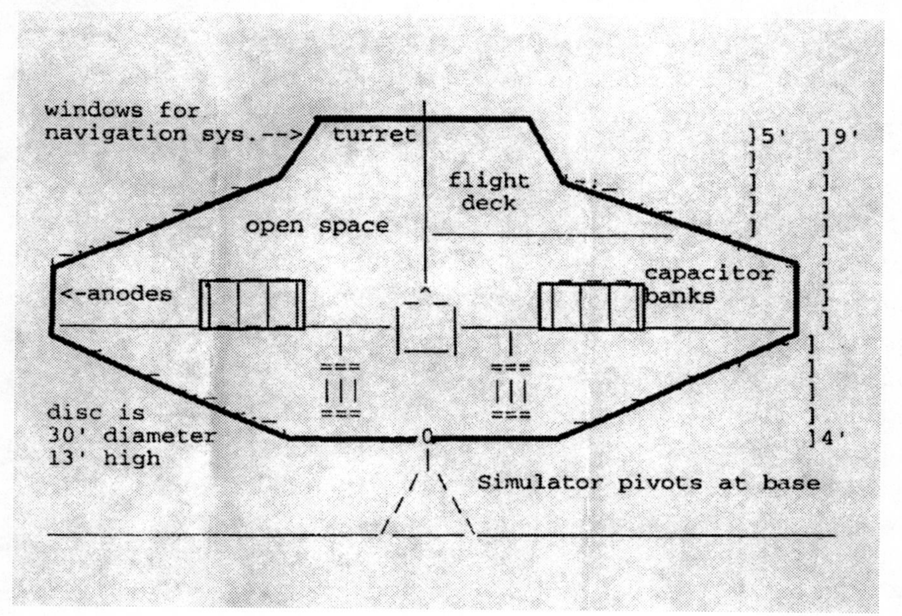

"The characteristics of this thing are very critical! When we completed the first one, each one of us that were involved in the design were required to be there for startup. I'd like to explain a little of how this thing was put into the building: It was actually on a pedestal. The pedestal was a large shaft with a large ball underneath, and the cross braces maybe 30 inches. And you could actually manipulate it. You could pitch or yaw this thing, to get the effects of ship movement."

Ques.: The simulator itself—was it tethered to the ground, or was it like an Air Force simulator, where you're sitting in a cockpit and have a big screen in front of you? What was it like when you got in? For instance, I've been in simulators, and you have a sensation of flying. What was the sensation of working the disc simulator?

B.U.: You had a sensation of the gravitation effects. You had the sensation of being raised on a large shaft. The ball it pivoted on was approximately 36 inches in diameter, highly polished. This shaft was like these lifts, it could pull that shaft up.

"When you walked in, it was actually flat, flush with the concrete [floor]. When they cranked the simulator up, to start it up—first, they raised it after you got in there, and they gave you the OK to go. It raised so you could make the necessary maneuvers of tilting, or pitching and yaw.

"The controls were all on the outside; all of the equipment for this particular model simulator was all on the outside. There were people on the outside of it who were to shut [it] down in case something happened, in case of emergency. Part of the emergency was that inside this [type] craft, in all of them as a matter of fact, it built its own gravitational area or effect: so you wouldn't know whether you were upside down, tilted or whatever which way you were. As far as you were concerned in there, you were sittin' straight up! It takes a while to acclimate. You just can't...for example: the people that we put in there for start-up...it was 3–4 weeks before we could get accustomed to the effects of the gravity inside that ship. You could take a person,

S4 Test - 300mm F 2.8 lens 2x adapter, Kodak 1000 film, 1992

without giving him the training, put 'em in the ship and if you said 'pick up that flashlight,' he would see the flashlight. But it would be a long while before he'll pick it up! It might take him two minutes, by the time he moved his arm over to pick it up; because of the effects of the gravity that was in [there].

"The reason [pilots] had to be trained is because flying discs—flying saucers—*they can't fly and turn like a normal aircraft.* Essentially, they stop; they have to stop. Say you're going south and you turn to go west. The ship will generally have to slow down, come to a complete halt. Power down. Make their computations of direction in the computer; the computer will turn the ship. Then, they can start to go west or east. So, it's not like flying a breather aircraft."

Ques.: Did you have a haze or blue color, or tint around it?

B.U.: Regarding the ionization: there were protruding rods, ionization rods around the ship. If you were outside you could see it. It created an atmosphere, is what it did; an ionization effect, like the ionosphere we have above the earth. Actually that's like a planet.

"The other thing about them—they can't use them as tactical weapons...for any war. They can be used as surveillance, [if] they have lasers, that won't affect the ship itself. Basically, the construction of this must be geometrically sound; perfectly done as far as dimensions are concerned. All of the items in there are specifically put in the correct spot. If it's built for 4 people, that's all it will fit—4 people. You can't fly it with 5; you may fly it, but not according to the original design. So, it is not something...[e.g.] say we wanted the United States to start using flying discs to go over and beat the hell out of Saddam Hussein, or his nuclear power plants, and manufacturing plants—it's impossible. They can use it for surveillance, or transporting people; and those ships are larger, they're 50 foot in diameter. Basically they're not used as combat vessels.

"I had a couple people, a few years ago now, give me this book—and this is supposed to be confidential—but this [book] was put out by Scalar Composites, Inc.

"There were 3 or 4 [people], and Burt Rutan is the person that owns the facility. And they had this guy named Gordon Novel, and Hal Puthoff I'm sure some of you have heard. They gave me this document which explained what they had planned.

"What their plan was—building a flying saucer ['Ufoil'; *The Phoenix Project*] for commercial reasons. And of course, as I just told you as far as the design is concerned—you just can't automatically just put anything [into the craft]. You know, when a number of people design something like this without having it *exactly* the way it's supposed to be done—with no mistakes dimensionally.

"These people, of course, they had their basic design. And I said OK, I'll provide some information to you after going over this particular document.

"What they had designed was something that would weigh about 32 ton. The capacitors inside the ship would weigh 30 ton before they're finished with it. Plus, they had some other problems: they weren't sure of the composite requirements in the shell of the craft [Scalar Composites, Inc.]. The shell of the [S–4] craft is 6 inches thick, and it's made up of several different minerals—boron one of them—most of those elements are used in building transducers and other electronic functions."

**Editor's Note:* This personal encounter recounted by Mr. Uhouse is a synchronistic, and wholly unanticipated verification of independently derived correlations. In a totally unrelated ufological study, an association of professionals were identified as members of a UFO-technology consultancy group: The Advanced Theoretical Physics Working Group. Identified member/associates, as Uhouse named some of them, were Col. John Alexander, Gordon Novel, Dr. Harold Puthoff, Dr. Robert M. Wood (see J. Alexander in Dr. Wood's documentary, *"The Secret"*), Burt Rutan, Jack Houck, Col. Philip J. Corso (member of earlier incarnation of group), Maj. Ed Dames (temporary), C.B. Scott Jones, Dr. Bernard Haisch, and Dean Radin, Ph.D. John Alexander disclosed to researcher Melinda Leslie that he had been an affiliated director of the A.T.P.W.G. He appears in the Uhouse account. And he's a witness to this solicitation of Mr. Uhouse, occurring at his home, in an ostensibly entrepreneurial R&D scenario. Whereas, the apparent expertise of Uhouse could prove invaluable; sort of a ground breaking stab at getting in on the ground floor, at the fringe of a new frontier: the paradigmatic transformation of technological evolution!

Overly ambitious—as ventures go—perhaps; albeit that the disc-frontier has already been claimed by the corpocracy, according to Uhouse. So, the "clearance"-free motives of the A.T.P.W.G. may yet be dubious at best, given the pattern of continuity among the military-industrial/Intel careers held by the members. Positioned, as many of these members still are, at the professional-fringe of the UFO community—their role may also be that of social-acclimation facilitators and monitors of the ufological Trojan Horse within society today; even if they are being "used" in this way by their contacts, and by default. I.e., the question is—since Bill Uhouse was a rather obscure figure—how did the A.T.P.W.G. members locate him a few years prior to Uhouse publicly volunteering their solicitation? And how were Intel-personnel authorizations made for vetting Uhouse's expertise file? The *Scalar Composites* interest in Uhouse also lends credibility to his account.

As a correlation-set, all this is crucial to the basic process of discerning credibility, and validity, to the ET/UFO phenomena as an issue of public policy significance. And lending further continuity to Uhouse, as a knowledge resource in this regard, is a second set of correlations. Converging upon the interests of the Scalar Composites group above, are similar data regarding the objectives of a Las Vegas philanthropic foundation: Robert Bigelow's National Institute for Discovery Science (now defunct). The two most prominent figures in the A.T.P.W.G./Scalar Composites nexus were Dr. Harold Puthoff and Col. John B. Alexander, Ph.D. They both performed additional functions within the N.I.D.S. foundation.

In Feb. 1995, a UFO lecture producer quoted Uhouse to me, resulting from a phoned invitation for him to speak. Uhouse said he had written a 16,000 word account of his career in four parts. He was quoted having sent the first part to the Bigelow foundation. Uhouse was a good artist, and he claimed to have redrawn the simulator schematics, interior/exterior, from memory. The remaining text and illustrations were ready to be sent, when his security monitors came to intercept the rest, objecting to his disclosure of certain details.

Bigelow founded N.I.D.S. in the pursuit of scientifically understanding parapsychology, paranormal and ET phenomena (these phenomena are all related of course). Thus, the value of employing Puthoff and Alexander at N.I.D.S. is their career based expertise in Para-Physics (psi) or parapsychological lab work. The career-link with virtually all of the A.T.P.W.G. members is this very same Para-Physics expertise. As Uhouse describes—the power-field effects on disc-craft pilots requires unique brain-reorientation, and acclimation. And if a private development firm (e.g. Scalar Composites, or perhaps even Robert Bigelow's company, Bigelow Aerospace) wanted to develop disc-craft, the simulation-test training would necessitate Para-Physics expertise. Perhaps mutual verification and sharing of how to understand the paraphysical/parapsychological applications—found in certain exotic propulsion systems—is why it was reported that N.I.D.S. had a quid pro quo arrangement with researchers among the NASA Breakthrough Propulsion Physics team and Astro-biologists.

Of psychological enhancement in training, John Alexander wrote his book, *The Warrior's Edge*. In his book, Alexander says: "Human willpower and human concentration affect performance more than any other single factor." Skilled application of Alexander's techniques mean, "using the tools of science to reveal the secrets hidden within the human mind." These, they call *psi*.

Thus, the correlations of interest in disc-craft systems, combined with scientific research into ET/paranormal phenomena—adds up to a need for personnel with expertise in such respective disciplines: i.e., the group soliciting Uhouse. It makes sense. And more importantly, the military, industrial and Intelligence backgrounds of these associates lend credibility to the Uhouse proposition that such R&D is a valid realm of investment.

Astute readers may recognize a correlation between the above personnel and their earlier incarnation: the *Aviary*. This third collective has been unofficially avowed by fellow associates Robert Collins and Richard Doty, in their 2005 book, *Exempt From Disclosure*.

An ethics censure ensued in 1989, when *MJ12* authenticator Bill Moore conceded his liaison to 12 *Aviary* Intel-ops members...

The interview continues...

B.U.: "Anyway, they [Puthoff, Rutan, Novel, et al.] gave me this document. I reviewed it, and we had a secondary meeting. I looked at it and I told them what my position was; my problem with it. They actually wanted to build it, as a matter of fact they had a check for 250 grand; certified check made out to this name, to maybe get into the act of going down to Edwards/Scalar Composites to start building the craft itself. And of course I had to refuse, as the $250,000...compared to the way I'm situated right now in retirement...I don't think they could match it. I don't get my retirement from the U.S. Government, I get it from elsewhere!"

Ques.: Who else; was John Alexander a member of that group also?

B.U.: John Alexander was with the group, yes.

Ques.: Who else, Gordon Novel?

B.U.: Hal Puthoff, Gordon Novel, Burt Rutan. And then a U.N.L.V. professor [Dean Radin, Ph.D. of the Institute of Noetic Sciences]. He was the smartest one there, I thought. [Burt Rutan thus returning to the arena of conformable space strategy, wins the *X Prize,* and the rest is prosaic history].

"Now, I want to say it's not the U.S. Government. They are U.S.; I mean, these are people from the U.S. that form this ['Group']. And there are a lot of companies that are involved, in an interface type of way, and they supply people. The thing about this, I call it *The Satellite Government;* which they do have essentially a government that's in a box that's separate from the U.S. Government. Although, they have interfaces with the various military groups and the Navy, the Air Force, and I'm sure even the Army.

Governing the Management

"Now, I want to get a little bit into what's out here at Area 51, and what's happened there recently. Getting back to...I'm gonna say *Satellite Government.* I'm going to say *my boss* was very upset when I said, and when I first brought it up was at that Friendship Campout meeting, that I indicated you know, I worked for a *Satellite Government,* not the U.S. Government! And maybe when I'm through here you'll understand what I'm saying.

"Out here, at Area 51, for many, many years we had captive contractors such as *E.G. & G.* construction company, plus *Wackenhut Security* and a few others. Anyway, what they had done, they've taken all those companies and they [the Group] sacked them; the contracts were ended. I think probably because of what's happening...about what is gonna happen in the United States; what's gonna happen in the world over the next several years."

**Editor's note:* Given the intensity of an escalating controversy due to Congressional passage of *The Intelligence Reform and Terrorism Prevention Act of 2004,* plus the re-election of George W. Bush, occurring at the time of this writing, it appears Mr. Uhouse was prophetic regarding "what's gonna happen...over the next several years" and enforcing civil liberties infringements.

B.U.: "And that is, whether it's going to be *one,* or is it going to still be split up, as far as individual countries are concerned? I think it will still be countries.

"But what's happened out here: there's a company, privately owned, by the name of *Bechtel.* They have a separate [subsidiary] corporation here that's just a Nevada corporation: *Bechtel-Nevada Corporation.* They are responsible for the whole site.

**Editor's Note:* In the interest of comprehensive validation, Uhouse reckons Bechtel to wield unprecedented power and influence for a corporation. This was proven in a controversy where Bechtel was just as unlikely a player as it seems Bechtel is playing in the *S–4* scenario. The case in point is how the Catholic church conspired to suppress an accurate understanding of the *Dead Sea Scrolls*. The content of these scrolls has proven to radically contradict, and historically impugn, the popular myth as to how Christianity of today mutated from the original "early Church."

In order to suppress popular or uninitiated knowledge of these historical/theological implications, the Vatican conspired in denying anyone access to read the scrolls, who were not loyal to Church doctrine. On September 22, 1991, Betty Bechtel of *Bechtel Corp.* initiated an "electrifying revelation." She entrusted "a complete set of photographs of all unpublished scroll material" to the Huntington library in California, thus, subverting the Vatican monopoly on politically correct interpretations. Synchronistically, since 1947, the scrolls had been politically co-opted by the Church. But, as stated in *The Dead Sea Scrolls Deception* by Baigent & Leigh, the photos of all unpublished scrolls were surreptitiously released "by Betty Bechtel of Bechtel Corporation, who had commissioned them around 1961." What kind of power and influence secured these "commissioned" photos? How did Bechtel achieve that which the most prestigious, world class scholars and university theologians failed to do for 44 years? What is the principle, and why?

B.U.: "The funniest thing about all of this is that there's two other companies: Johnson Controls and Lockheed. Lockheed being mostly involved with aircraft frames [etc.], but not the propulsion system that makes the plane or the disc work.

"Johnson Controls basically has been involved in the part that makes the disc work. If you want to call it a reactor, that's fine. I'll go into some of that in a minute. But it's the propulsive system for a disc.

"Now, I'm going to explain why these companies exist: Johnson Controls was a small company in 1967 (they had 25–30 people). And this is going to astound: they started off as an instrumentation [company], *Johnson Instrumentation and Controls,* that's what it's called. They started out in Pennsylvania.

"In that group they had two people, and I'm not going to identify the names. They were two people to get this company in a position...to go out here and take this over.

"They *[Johnson Controls]* do a lot of things. They've built them up so big. They're not just in the U.S. They make things in Europe—plastics. A lot of you know the Diehard battery—that *Sears* produced—was put out by Johnson Controls. They really had a hedge on the battery. The reason why—because of certain technology...given to Johnson Controls as part of this team, part of this organization.

"As new things come out—as a result of this *scientific transfer of technology,* as a result of our building these discs now—they are given to certain companies: General Electric, Johnson Controls, Lockheed has got a lot of it. You got other companies that are given certain things to produce, to build and sell on the world market.

"Of course, one other thing I'd like to point out, a lot of our old technologies are given away. We're giving away patents to China, a lot of things to Japan, and Third World nations, so maybe they could develop and come up to some level in society. I don't know what it is, but we're giving a lot of our old technology away.

"*BUT*—we do have, uh, an ace in the hole! That's what we've learned from the *G.E.T.s;* the *G.E.T.s*, meaning Gray Extraterrestrials. The gentleman...up [speaking] before me talked about Grays and a few others. And yes, I'm sure there's some. I don't know those. I'm not familiar with *those* Grays, for example, or any other types they've indicated like reptilians and that type of thing.

"The Grays involved in the *transfer of technology,* for the particular disc we were building, were from this particular group; from the

Reticulum Group, which is supposed to be 39 [37?] light years away from here. [See *An Alien Harvest* by Linda M. Howe, p. 150.] I just wanted to explain some of the things that are happening.

"So, now we have Bechtel Corp. running the whole show. If you look at any newspapers in Las Vegas, they only mention the Nuclear Test Site; that's all anybody thinks about. But there are more things *out there* than just nuclear work.

"They've had many, many years to dig these tunnels, and put their secret facilities underground, rather than on top of ground. That's been planned many, many years ago. The funny part of it is—I'll just let you in on one thing—Bechtel is the basic designer of the tunneling operation. They may not have had the people there, but they did the form and design. If anybody here is from Europe, they'll know who managed the *Chunnel* over there—it was Bechtel. Bechtel is one of the key members, as far as I'm concerned now, in this *Satellite Government Group.* Even though Johnson Controls is on the stock market, I think what they're trying to do is pull away from anything to do with the stock market, commodity exchange [etc.], so that private companies out here at the energy facilities...so everything can be kept more secret on any project that might come up.

"It's like, if you've got a company like General Electric, although there are G.E. people involved, they're separated. These [management] people may pull them [in], like they pulled me and my other friends. They may pull them and finally you will be working with them, as a result of our going through all this clearance, and being given the bucks to do it, and making your life comfortable. That's what they've done for me. I won't have to worry til the day I die; I'm in good shape. A lot of people have to worry today because of what's happening outside: Companies downsizing and things happening in this country. They want to make everybody...below the average income person today; they want to make them a peon. That's what I believe, and that's probably gonna come true!

"Speaking about that, there's a point: People talk about the right wing, left wing, militias, etc. I really think that before [conflict] happens, the *Satellite Government* is waiting and standing by to see what happens politically, both here and abroad, before they assume control.

"Now, what you're going to have, you're not going to have a Constitution. You're going to have a dictatorship! Whether it is going to be by the military or by big bankers, whatever, but I don't think it will be by big bankers. I think it's going to be [military], even though the ones [I worked with from] the military never wore uniforms. I knew who they were and who they worked for, and yes, some of them worked for the military. They were *from* the military, and put in the *Satellite Group,* but *still* an officer of that group, in some ranking order.

[Regarding personnel choices] "One of the things about the people in security: I know somebody out here, in the back of their mind is thinking: 'Well just what kind of people are they hiring?' And maybe they know of Bob Lazar, and they say 'How come Bob Lazar got in there in 8 days, or 30 days [etc.]?' Well, I want to explain one thing: Lazar is a *Hungarian!* I bet that doesn't mean much to anybody, I'm sure, but one of the planners—involved way back—and several people under him were all Hungarian. The reason why I'm saying this: When I worked on a newly drawn design, and we had drawings given to us from other sections of *The Group,* there was always a notation on it identifying the level of secrecy; and other pertinent information they necessarily wouldn't want you to know. [The secrecy code was in Hungarian].

"Number one, I don't know Hungarian. If you ever look at Hungarian and tried to understand it, you wouldn't be able to. Hungarian is no different than Indian language here in the U.S. where, during WWII [against] the Germans, they used Indian codes to confuse the Germans. The same goes for Hungarian. There's no language in Europe that's compatible with Hungarian. The closest one is Finnish [Linguistically, the European languages of Hungarian, Finnish and Basque are not Indo-European in origin; nor of any other known language group. Zecharia Sitchin explains, they are rooted in Sumerian, and in turn with the Nefilim: "sons of the gods who came to earth from the heavens"].

"As a matter of fact, I used to see it, and I asked my boss what that was? He says, 'That's Hungarian classification on the drawing.' OK! That's all I needed to know [or, had a 'need to know']. Right! It's none of my business!

"And, I'm going back to Lazar again—the reason why he was hired: he's a Hungarian! One other point, they only hired W.A.S.P. people. White, Anglo, Saxon (not necessarily Protestant). You could be a Catholic. They want to make sure they had all Europeans. There were no Japanese, no Chinese, none of the other nationalities that we have around the earth; that's the way *The Group* is.

"I will mention this also—at the other meeting I'm sure I did. Maybe that's why I got hit? But I'm just thinking about it. They wouldn't [describe] what I really said. They just told me to be careful what I talk about. I'm bringing that up to see what happens again, [with] this new information I'm going to provide today.

"OK, here's the way it went generally: I got out of the service, and I get the job at *Wright-Patterson.* I applied for a job and got it, at the time *Wright-Patterson* was doing testing on the *F-89,* or modifications on several other things. They had the *Northrop Wing* out there; I flew in the *F-89, F-102* and several others.

"From that point on, I was approached. I was approached by a colonel. He asked me specifically if I'd like to go out to the special projects. One of the projects was: they wanted to take this metal and join it without any welding or any glue or anything else. I told this story before, but that was one of the projects I was assigned to. And right now there is one plane that's flying, the *SR-71,* their fuel tanks have these particular joints that expand and contract and still stay together, when it's fully loaded with fuel. That was one of the key things. I worked on that project, that's how we got started.

"So, after doing these various different projects—these particular projects were special in the respect they were never done before—my friend and [I] said, 'hey, maybe we'll get to work on something really big one day.' We went on several different ones. Finally, they said: 'We would like you to come to a place in New Mexico; spend a few days and talk to some people there.' And we did. Prior to going there we also worked at various simulator companies, *Link Aviation* of Binghamton, New York being one, and a few others; they were small tours. They wanted us to get the experience on [flight] simulators and I'm thinking: 'why', you know? I could go out and work for Northrop or Link Aviation on a simulator.

"That's basically how we got brought into the program. Eventually, not necessarily that particular day; like I said, it was 4 1/2 years before we knew what we were working on!

"When they showed us this first monstrous assembly—it was a big round unit—we didn't know what it was. Just a big round thing: like two woks upside down. The unit was monstrous, cables running in and out. They took us around and showed [us], trying to explain what it was supposed to do. Of course, we didn't know what the hell it was supposed to do.

"My job was to be avionics: I was one of the design-mechanical engineers, [doing] redesign of the instrumentation so an average human could fly a disc; instead of having an alien along with you to give you instructions on how to fly it. This took *some* time! I worked on this thing for 38 years! But we did more than one in that 38 years. The most difficult part was finishing the first one. Of course, when they were building the simulator, they were building the one they would fly right along-side of it.

"It was already a 'black project' when I got on it, so somebody had a plan at the time. They knew what they were gonna do, they just needed the people. In those days, even today, you don't just pick people off the street without *knowing* them, understanding them, to make sure that psychologically they're going to stay with you. For example, [to] not commit suicide or run off somewhere. 'Cause, they'll get you! After a while, not in any threatening way, but they'll get you, somehow!

All of the Above May Be True

"[The] flying-disc simulator; now, let's get into that. The first [craft] we looked at was apparently some craft they picked up from somewhere, I don't know where. I don't know anything about the Roswell crash. I can only relate this to the Kingman [Arizona] crash. That's the one they picked up, and that's the one we re-designed to fly; was the one that landed at Kingman. The reason why we picked that one out to fly: it was very similar [to our design options], but less complicated. It didn't have a lot of things that you needed to really get into this thing and get experience and fly the damn thing! So, what they wanted to do was take the easiest one first, which they did.

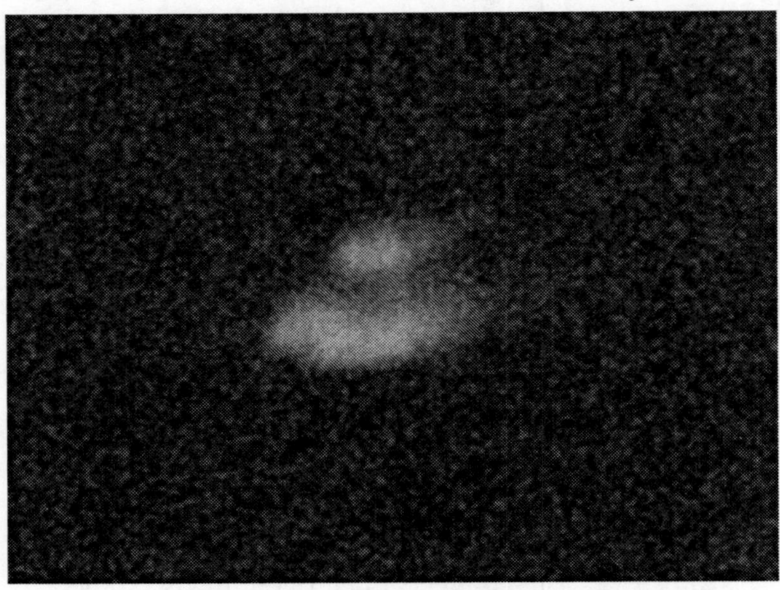

Photo by Gary Schultz, taken 2/28/90, at the "black mailbox", f 2.8, 1600 ASA film

"Now, if you looked at the Lazar disc, it was exactly the same: two woks upside down. Except it was 3 to 4 feet between these woks. The reason why? Because we got tall people. We needed room to get around, we needed room for our avionics. Our avionics were much larger. In those days, in the 50s, we did have some simplicity, some electronics reduced in size, but not all the gyros, etc. The flight indicators were still large; they were big for that type of aircraft. You gotta understand this thing is only 10 meters in diameter; we raised it apart 3 feet, to put more equipment in it. I'm going to tell you how we operated it; I don't know anything about the reactor.

"I was responsible for the avionics part of it, nothing else. The operation of [the simulator] was all together different than was in the actual thing that flew the actual craft. For example, I've never even seen the actual craft run. We were on the simulator, we were very well compartmentalized. And we had our jobs to do...we did them. And it just wasn't done [say, as] at White Sands. There's a special facility down there that was made probably better than out here, and also at Los Alamos.

"The actual construction and design of the disc-simulator took approximately 16 years (from the time I was aware of it). The Kingman Saucer, the one they picked up, was in 1953. From that disc they had four occupants, which were transported later, out here at Area 51.

**Editor's note:* A logically plausible detail, outlined below, could be crucial to disc re-design protocols Uhouse alleges. It is included as data for verification.

About the time Uhouse spoke publicly (1996, '97, '98), an abductee/collaborator of mine received corroborating data from *another* alleged "insider." This other insider was retired, still consulting under oath of secrecy, yet willing to converse off the record. He described having direct involvement with *Project Blue Book Report 13* for about 3 years after WWII; he claimed propulsion system knowledge. Due to the abductee's own experience, a mutual reinforcement ensued. She felt she learned technical specifics about craft propulsion from ETs in her experience. Oddly, her specifics matched verbatim the insider's understanding, due to his direct involvement. After volunteering just a few clarifying details, the abductee exchanged on a topic which, in her experience, was too technical to share without mutual understanding. This synchronicity surprised her. It validated her experience, while confirming her credibility to him regarding close contact experience. Specifics, pertinent to the Uhouse account, were alleged to have appeared in *Blue Book 13* autopsy reports. According to this insider, principles of propulsion/craft re-design were born out of crucial differences between ET and human physiology. Questions about the propulsion/craft which made no sense were answered by the autopsies. We couldn't move a vehicle like ETs do, since our anatomy is not built like theirs. How we duplicated that had to allow for anatomical differences. The abductee felt the insider's info confirmed details she perceived in her ET experience!

B.U.: "Now, back in the 50s, there was very little traffic around here. The disc was brought in from the eastside of the Calico Mts., north of Kingman. That's where that particular craft landed. It was close to the Army airfield at that particular point. [See *An Alien Harvest* by Linda M. Howe, p. 149.]

"In the operation, I had spent some time—after the simulator became operational—so they could test it, for example in the start up. We went through the whole start-up phase and had to perform to certain specifications. I was asked this question: 'Was the door open [during operation] when they were inside?' I said, 'no.' It wouldn't operate [with the door] open, you had to have a sealed envelope. Once you got inside you would identify very different feelings than...being in this room. As a matter of fact, you had to go to many, many sessions, like to psychologists. People that are identifying [for] the individual, you know, how you are going to feel when you're in this particular craft; sitting down, standing up or whatever. I'd like to explain the [feeling] that's most prominent: when you go in, and the pilot says 'crank it up' (you'd be sitting at one of the consoles)you're going to feel *pressure,* all over your body; as a matter of fact it'd be difficult to move your head. Now, if you didn't know that, it would take you quite awhile before you'd be able to raise your hand. You wouldn't know! You could, but some pressure was just pushing you. The reason for that, what it's doing is [emanating a field-density of electric charge from beneath the cabin, where a bank of capacitors release a pulsed field]...used in oscillation of what they call a flux-plate.

"Now, I'm gonna say one other thing to try and make sure everybody understands: The disc is no different than a planet. It's got a north pole and a south pole. There's no front, there's no back; there's no top, there's no bottom. If you're flying upside down in the disc, as far as you are concerned, you are right side up. I'll give sort of [an analogy]: Where's the top of earth? Everybody says the top is the North Pole but that's not true. There's no top, there's no bottom. [A quantum insight may also be applied with this planetary analogy to disc-craft. The operational craft may require its propulsive powerfield to be harmonically *tuned in phase resonance* with the magnetospheric earthfield, for efficiency. Uhouse commented on this 'resonance' factor in private discussion].

"The only way they could vector the disc was to design this criterion into the particular one we developed; [utilizing] our science, our know-how, as far as the disc was concerned. So, at least the pilots could—until they got enough experience flying out of the atmosphere—get really comfortable with it; knowing all of the phenomena that's going to happen to you inside that ship while you're flying it.

"Now, they've asked me—'Was there a restroom in there?' No! I don't know where the restroom was; we didn't put one in. Whether they had one on a real one or not, I don't know. It was 10 meters in diameter; there's only so much you can put in this particular ship. If you put certain things in there that aren't supposed to be in there, you could end up in trouble. There were requirements and criteria defined, what we called *FOGIT.* That's gravitational technology; sort of a specification on what you can put in and what materials you can use. A lot of the materials that we used were exotic. A lot of the materials we used in [the simulator] were the same as we used on a ship, so we could get the same response and characteristics when we energized it for doing a simulation run.

"So, you've got to remember: every one you see up there, if you see one, it's got a north and south. If it didn't, they wouldn't know how to vector it: Which way do I go, left or right? How am I gonna get there? Each time they do that vectoring, it has to be calibrated. Calibration is done in a couple of seconds—boom! And you'll go in a different way. A lot of people who see them fly and say 'Boy, it made a sharp right turn, sharp left turn'—that's because they wanted to go, [then] recalibrated [their direction].

"The capacitors pulsed the flex-plate, which was on the bottom of the ship. Under this flux-plate were the gravitational type amplifiers (Although we didn't use them to the fullest extent, because we didn't have a reactor). Plus, we didn't have the heat. We wanted to do it electrically [electro-gravitically], which we did. It took awhile.

Ques.: How fast would the saucer go?

B.U.: The speed on that was, I think about mach 26 [approx. 20,000 mph]. It had 6 capacitors inside this particular simulator, each one at a million volts, for the power output...you see, the skin of the ship is 6 inches thick, and it's laminated with 16 shells of material. So, in effect you got a big transducer: I don't know what materials went into it because that isn't my area of responsibility or wasn't my "need to know."

Ques.: You mentioned a commercial reason for this. What would be the purpose for having a ship like this?

B.U.: It'd be interesting to fly them! The problem is—let's say I have a 160 passenger flying disc, right? And I'm gonna fly from the

U.S. to Europe. I've gotta give them about three weeks training before I can put them in the ship. That's going to be expensive! It's not real practical, no.

Ques.: What was the power source?

B.U.: On the simulator it was capacitors. Stan Deyo, from Australia, indicated that they had a squadron of saucers in Australia, South Africa, and of course they have one here.

Ques.: What year did you start on this?

B.U.: I started work on this in 1953.

Ques.: Did you ever see a working model or ship?

B.U.: Not a ship; a simulator only. You see, this particular organization called *The Company,* as far as I'm concerned, they're a *Satellite Government,* they're an entity of their own. They play people, or they hire people, or they take people from the average population. For example, [a certain person], if they wanted him, they'd be investigating him several years before they even talked to him—say 3 to 4 years. Like this Col. came up to me and said, 'I'd like to talk to you about a different kind of a job.' They look into your past. And you don't know they're looking into it, but they're checking you out.

"If they are satisfied with you, they'll put you up. Once an employee of the organization you're forever an employee! You get paid by cash!

Ques.: Is that [simulator] you worked on still in existence?

B.U.: That one was shipped out here [S–4]. I haven't been out here for years, since 1993. I left the program in '89, but was recalled back, and I was back in two years.

Indoctrination Plus Acclimation GETs the Next Generation

"You see, the problem is, they got a big problem: the guys like me that are leaving...like a friend of mine, one of the first pilots that actually trained in the simulator that I know, is retiring this year [1997]. But we need more people, you know? There's several people like me out there. But somehow they're not in the same situation I'm in, and I guess they got me locked in for several different reasons, into Las Vegas...here. If anybody's looked in the phone book in Nevada, I'm not in there anymore. It used to say there was a Bill Uhouse who lived in Nevada. There isn't any more. He's gone!

Ques.: The actual craft, when it's flown, it's capable of cruising around the planet. Is it able to go on to the moon...is it able to go out in space to a certain degree?

B.U.: Yes! I'm sure we're going back and forth to the moon. I'm almost positive of that; to Mars, I don't know.

"You know, they talk about the different ships they are building—I'm sure the *Satellite Government* is constructing the *triangles...triangular craft.* The reason why they put [it] in the triangle shape is so they can charter people, charter supplies; have the necessary facilities for people [toilet], and make it more like a plane. With this [disc], it's impossible. And I saw one of the triangles. Out here as a matter of fact; two years ago when I was out at 3:30 A.M. I just came across it when I was down at S–4 area. But I've been getting e-mails from England on triangular craft. Several people are seeing them back east. So, I'm sure they're constructing them. I don't think they're alien, I think they're human made—by Anglo-Saxons!

**Editor's Note:* Converging in the following account are a number of correlations, adding continuity to the Uhouse reference above. In her book *Diary of an Abduction,* Angela Thompson Smith, Ph.D. recounts a field investigation she conducted. As a research liaison for Robert Bigelow's *N.I.D.S.* foundation, Smith and Bigelow interviewed local people in "some small towns north of Las Vegas," Nov. 1992: "There was a local cafe owner and her husband who reported that...they saw a huge, black, triangular craft fly silently down the canyon and over the town. The craft was accompanied by two military helicopters. The sighting was repeated every night at the same time, with such regularity that the woman and her husband brought out lawn chairs every evening to watch the flights. Other townspeople had seen the same black, triangular craft flying low over the valleys between Caliente and Pioche. We heard from another woman who reported seeing a similar black, triangular craft flying low over Caliente. She said that it was as big as the hotel it was flying over. She was with a group of three other people when the sighting occurred."

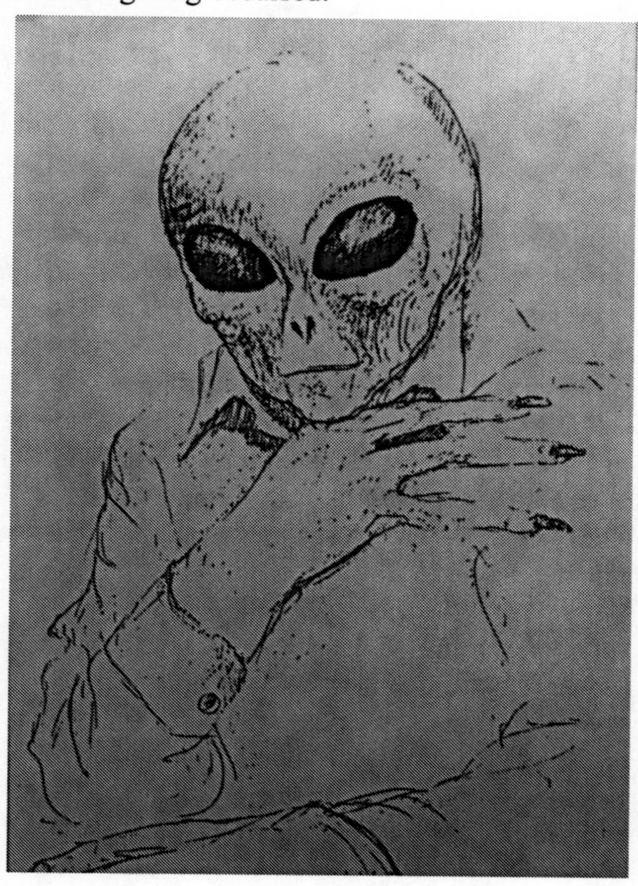

EBE Drawing by Bill Uhouse

B.U.: "I'll bring up the matter of tunnels: Yeah, there's a lot of tunnels out here. For example on the Mercury side, they have a real large tunnel: about 60 to 70 feet [wide]. And in these tunnels they got these cells—as far as you can see. And in addition to test cells, they have other types of rooms [where] they do various work: not necessarily nuclear—It could be biological and a lot of other things. A previous speaker showed photos of these places...opening up out at Edwards Air Force Base, in California. That's the same thing we have out here at Area 51; and you can go from Area 51 over to Papoose Lake and several other places.

"It's basically that people fly in from California; Langley AFB [etc.]. At McCarran [Vegas airport] they have their people go and [fly] them to work, out here. The funny thing about workin' out here is: if you're an engineer, doing various different designs and testing etc., you might spend 2 to 3 weeks out there without ever going home. They take care of you: the food's out there, the drinks are out there. It's better than any Officer's Club...just fine. The food *is*—really. You can have lobster every night! But, there are [tunnels], more than just from Area 51 to S–4 if you go back into the mountains further. You'll probably find—if you can get in there—various different kinds [to] various places. You could enter [them] at various locations. They may put you in a bus to go there: It will be blacked out, so you won't know what direction you're going, or where they're gonna let you off. When you finish your work you come back the same way. That's the way it is out here, and that's the way it is at all these facilities.

Ques.: You mentioned there were 4 entities from the Arizona [crash]?

B.U.: I'm saying there's six species here. That's what I understand, six species of Grays. And they call them, as one, the G.E.T.s.

Ques.: Is that various sizes?

B.U.: You know, they talk about these little guys, but I don't know the little guys. I know [them at] 5 feet 4, 5 feet 6.

Ques.: Did you work with them personally?

B.U.: I've met—this is part of the indoctrination and security—I met a gray, and he was a science translator, way back in the 50's. He was the one that provided [assistance]. His name...we translated 'J-Rod.' But if you looked at it like I'm reading Hungarian, you wouldn't know what it was. There were other aliens. He was no different than anybody else, except he had different features.

Ques.: Did he speak through his mouth or telepathically to you?

B.U.: I'm gonna say this again—he spoke *in your voice.* Whether you hear it telepathically or not, you hear [it] in your voice. Oh, they can speak with their mouths, sure. I think I saw *his* move several times. These particular individuals, they're different than us in a lot of respects: Their features differ, as there are differences between Chinese and Americans and French and German and whatever.

"I've got to explain something: I enjoy listening to other people talk. For example Bob Dean and others who talk about the UFO phenomena.

"One of the things that kinda bothers me most: There was a panel that included Randolph Winters and Cal Korf that discussed the Billy Meier case. It came out in the panel that they mixed the technical portion of it—regarding the disc he saw and the Pleiadians—with religion! Jesus Christ and so forth. When you start mixing religion with the technical portion of it, of anything, immediately you're gonna get in trouble! People see flying discs—they do; and it's not that we're not flying them—we are; and it's not that aliens aren't flying them—they are! And they come in different shapes.

Ques.: What about the gray Reptilians we've heard about, in the underground bases?

B.U.: Let me tell you this—I'm gonna tell you about Reptilians. I'll tell you, *there aren't any. There are no Reptilian Aliens!* I don't know of any Reptilians, and besides, here I am and I'm gonna take a snake and clone it with me? You know what's gonna happen? One of us is gonna die. Now, this is science that I know, as I understand it. I've talked with biochemists I worked with for years. We've even tried that: We humans have tried to do that type of cloning, and it was unsuccessful. Now I don't want any of you to get upset about [me] sayin' there aren't any Reptilians. It's all in the mind of the *beholder!* It's like the question people ask me, 'Well, do they talk?' Sure, they talk! Hey, I'm up here telling you what I know to be what I understand, OK? I'm not giving you any story.

"And I'll give you another good example: When we went through the psychological process of being introduced to a gray EBE, there were maybe 30 some people in the group I was in. Do you know that only 16 of us passed the test and was allowed to meet them? Do you know why? Very simple! It's your mind, OK? It's your mind.

"I'm gonna try to make an example so some of you understand: Say, for example, I'm an engineering officer, and I'm thinking about engineering things, OK? Nothing else. And here comes this gray alien, and now we're thinking the same thing. Now, if something different came in, that I was in love with, and they looked different than me for some reason, and my mind all of a sudden says: 'Goddamn, he's a ghastly looking creature.' My mind thinks he's something else than what he is—you're looking at something else than what he is.

"Now, if this guy [alien] walked into me, and we were thinking in engineering terms, and all of a sudden I changed it to *girls*—forget it! So, it's having the same train of thought; the same mindset, [concentration].

Ques.: How long did the psychological process take?

B.U.: It wasn't anything that was fast, it was 3 to 4 months.

Questions of Acclimation

Ques.: This may be a very basic question, but why are [EBEs] here collaborating with us on engineering?

B.U.: Let me say this: There's a lot of things here that they want, for instance certain natural resources.

Ques.: Do they easily adapt to our atmosphere for life-support? Is their atmosphere identical to ours?

B.U.: Sure, sure—basically. Maybe our oxygen content is maybe a little more or less.

Ques.: And oxygen is their mainstay of life as it is for us?

B.U.: Well, like I said, they've probably been coming here for years. But we weren't really smart enough to understand what they were doing, 'til something happened.

Ques.: Is there any indication of a hierarchy among them; like is someone in charge?

B.U.: Yes, they do have a hierarchy. They have what you would call an understanding...and hey, it's no different than *here*, it's age!

Ques.: So did they give you any impression of their lifespan?

B.U.: Lifespan? Well, I understand some of these guys are like 250 to 300 years old. They have a real understanding as far as age is concerned.

Ques.: I feel there may be an analogy between the *Philadelphia Experiment System* [of stealth] and a flying disc. Where, on a system like [a disc], when they powered it up, it should disappear. Is that the same technology making these craft disappear? Would it accomplish the same result? I've heard many people say, when the ship powers up, the light goes around the ship.

B.U.: But if you were on *top* you could see it. And seeing it from the bottom depends on the bottom, and how big the ship is. And the ion rods around there—that make the ionization field—makes that happen too. It would disappear to you, but it's still there. The time-warp element is altogether different.

Ques.: The Hungarian guys dealing with the propulsion info: were they also involved in the medical and psychological aspects?

B.U.: No, not in that group I was in. I've been asked that question, Gene Huff asked me, and a couple other people. I said 'no, I did know a bioengineer, but...'

Ques.: Would it stand to reason that if they're involved in the propulsion system and training of pilots—like you said—part of that training of pilots is major brainwashing, or bel...

B.U.: Yeah, it's *a reorientation,* is what it is.

Ques.: ...or belief-system change. But, that wouldn't say that if they're involved in propulsion, therefore they're gonna have to be involved in that psychological training of pilots? The Hungarians involved in possibly the psychological training as well?

B.U.: Ohh, you mean where I worked; I guess they were, yeah.

Ques.: Regarding [the] alien's appearance: what color of skin was it?

B.U.: It was pink-grayish. Not as gray as they depict them. Well, they look like they're alive and they got pink to their bodies—I mean, it maybe looks a little rough. They have to use a lotta lotion.

Ques.: Did you see the *Alien Autopsy* on television?

B.U.: I saw that thing. There is some film out there that's real—I mean, *actual.*

"This [*Alien Autopsy*] film was a real film: in other words, it was probably taken in 1947. But let me tell you how the Government does this: They'll say, 'you, Shawn, this gentlemen, and Bill and somebody else—you're photographers. You go to tent No. 1, you go to the hospital and view the autopsy,' and so forth and so on...Only one of those out of the 5 or 6 is the *actual* one!

"Before we go to lunch I wanted to bring up a couple things: Number one, gasoline prices! Gasoline prices are kinda high, right? It's all like I talked about: how they plan to get you [economically]. There are some people trying to formulate new gasoline, I know. California has done something. I don't know what's in it, but I know one item that's in it—it's *Boron!*

"When I was younger, I used to see, I think it was Sonoco Gas stations, you know, 'Our gasoline contains Boron,' right? The reason I'm saying this is because, the ETs, that is one of their tradeoffs: We give them boron. There are only a couple of places in the whole world that have boron. One is out here at Death Valley, the other one is near Edwards [AFB]. You might find some in South America. Those are the only two places you can get boron. Boron, boraxo, call it what you want. Now, you're brushing your teeth with it. I just want to tell you how important this is, ok?

"The other things containing boron? They're putting it in vitamins now. They are also taking boron and injecting it in humans to correct cancer. And they make airplane parts out of boron. I think one of the first planes they used it on was the F–111. You can make it hard, you can make it soft. There are so many different things you can do with boron. You can drink it; it's explosive. You can inject it into the fuel. It gives you more power for all kinds of concerns. But boron has so many good things about it, that's why the Grays want it.

"Maybe this is the only place they can get boron, but that's one of the things we were trading off.

Ques.: In addendum to the tradeoff with boron, are they not exchanging human bodies? Or is there some sort of thing of that nature? Do you know anything of that nature?

B.U.: I don't know anything of that nature, as far as trading human body parts or trading humans. I just don't know. You know, I'm like a guy...for example, people live in various different towns. They live in a location like Chicago. They live on a couple different streets; they got their movies there, they buy their groceries there; they work there and everything—and they *die* there! Right? Well, I've been sort of in that position with this *Group.*

Ques.: How many people live in the facility, in total; what would be your guesstimate, at S–4?

B.U.: It might be 20,000 total, interfacing [etc.], but the actual people out there? Probably 3,000. If you look at the planes and so forth, there are a lot of flights that come [in] straight from wherever. They come from Northern California, or Los Alamos or White Sands or that kind of thing.

"I'd like to say this: You know, they say out here they moved it [the operation at S–4] here, there. But nobody's that dumb that they would. I don't know how many billions, maybe a trillion [black-budget] dollars that's been spent out there—whatever they got totally in the facility. And move it anywhere else, you know? Just to rebuild or build a facility? It's a problem.

Ques.: How much did this cost?

B.U.: I'll tell you the best I can: probably several trillion dollars. There's things out there, you know, [People] say the place has been moved. They can't move it. They've got a D.C. generator out there;

it's the largest one on the planet. And it's impossible to build another one someplace else. Plus, they got other facilities underground. It's impossible to move it.

"Out There" Was Moved How and Where?

"The rumor of Area 51 moving to Utah or White Sands, and a couple of other places, Australia, etc. I've been out there, I *know* what's out there. And to expect them to remake what's in there today is practically an impossibility, and I'll explain why:

"There are two items they can't move: *One*—the largest D.C. electrical generator on this planet is right out there. They use two types of electricity: The direct current is used basically for the functions that they do in tests and scientific development out there; all done with direct current, not A.C. The A.C. is only used in the subordinate facilities, where, as an example of one, there's this big airstrip that everybody's looked at [in aerial photos].

"I'd like to say one other thing: *Area 51*—such as this airstrip and the housing of other facilities that are there—[is] only for the air-breathing aircraft. It has nothing to do with the EBE portion of [the base]. The actual discs that are constructed, and that they have out there *are not in Area 51!* They're on the other side of Papoose Mountain Range.

"A lot of people have said: 'Well we've taken photos from the air and we can't see anything.' And that's all part of democracy, but I'll explain what democracy is later. This is not a joke, really it's serious.

"The *other* items they have out there: Some of you people here have had an M.R.I.—Magnetic Resonance Imaging of a person. They have the largest [M.R.I. machine] on this planet, right out there. You can put a whole craft in there. With this you can back-engineer new items, or something you've never seen before—its internal mechanism. You can put it inside that [M.R.I.] machine and you can see every detail on

Aerial view of the S-4 site, from a declassified KH-4B reconnaissance satellite, Aug. 15, 1968

it. Maybe not in the minutest detail, like after you take it apart, but actually the basic stuff. This is not new. This particular [M.R.I.] was built back in the 70's. M.R.I. for people did not come out 'til maybe 1988–89.

"What they're trying to do with this out here, is kind of keep people away...from the area. I can't see any reason for telling people not to come here. They can see anything,... the [AF jets], or it could be a disc or some other object.

"And yes, there are aliens out there. I don't know how many, now.

"The *other* thing out there that's basically irreplaceable are the *cells* of the underground facilities. I know people've heard about the tunnels, etc. I'm not aware of the tunnels that go from here to Edwards [AFB] or anyplace else, but I know about the underground facilities that are out [at S–4] where the work is done. And the EBEs out here are there working with people like myself, on various different objects and subjects.

"The most important thing is, I want to explain is this: There may be other facilities that are coming together in Australia, and maybe elsewhere around the world. But, if you look at what's out here, it's impossible to put it all together and duplicate. Not by billions of dollars, but trillions of dollars that's been put in this facility. Just designing and *placing* and developing the equipment with all the designing and engineering required is gonna take many, many, many years. So, for example, if I'm gonna put it at White Sands, it's gonna take me 10, 15, 20 years to do it. And the people that are involved in doing this are gonna be a little bit upset! And those are the companies providing the money to get the new technologies to build whatever, any new thing that may come out of the research that's done here. For example: G.E. may provide that information; so they can build, depending on their expertise and their discipline and how much they're involved.

"This money they're putting in: it's not all black-budget [DOD money] that you hear about and that Congress has appropriated. A lot of this money is money from large companies like General Electric, Westinghouse, I.G. Farben in Germany (for chemicals, that type of thing), [Lockheed, Bechtel, Johnson Controls, Aerojet General, Hughes Aviation, Rocketdyne, Glenn L. Martin Co.]. They provide the bucks to pay guys like myself to come up with new processes or new ideas or new science developments.

"And really, if you think that we're at the end of science and technology—we're not! We're just beginning! So, I'm just trying to explain: all this you read in the newspaper, and that you read in some of the magazines—take it with a grain of salt!

"I don't know how many people have computers that are with us today. But there's a lot of info goin' over the net on this whole subject. And there's also one other person I want to mention and that's Stan Deyo, out of Australia. He claims to have worked for the *Secret Government,* or the *Satellite Government.* He has talked many times on the Art Bell Show, and there's a few things Stan said [that were] actual facts: The fact that they are building a facility in Australia—they are. But now, what they need to do, and what they are going to do...it won't necessarily be companies from the U.S. that will be supplying money. It *will* be companies—from England and Australia. They are *so far behind* [compared to] what they've done out here, and what they're doing out here; it's gonna take them years to catch up. But you know, generally we need a plan like that—to have a back-up. There needs to be a back-up in case something happens here, you know? Right here at Area 51. But I don't think anything's gonna happen that's too serious—I think it's gonna continue on. I think there's going to be a lull first, for the time being. I think in the next few years that there'll be more activity; that you people will enjoy coming out here.

Simulation vs. Operation

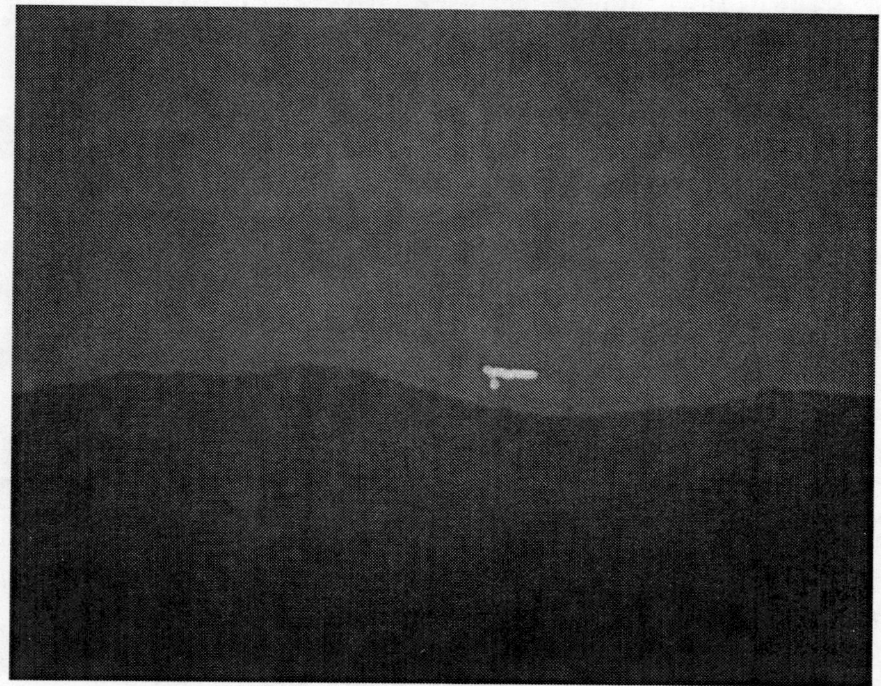

S4 Test - 1600 Color, 3 sec. shutter, 1992

Ques.: Where are these craft being flown to?

B.U.: I'd say there is two different ones: They have a craft that flies just in our [atmospheric] space. And they have craft that fly out.

Ques.: Where do they go?

B.U.: I don't know. Who knows? I don't.

Ques.: Who flies them?

B.U.: I'm saying that now we have trained humans flying them.

Ques.: Did you work on two different types of propulsion systems? Or did you know about two different propulsion systems? One would be for in our earth atmosphere, one would be out of our atmosphere?

B.U.: This is [for] earth's atmosphere. Now, let me say that the hull we built was 6 inches thick. You had two different plates: It had a special interior, like [when] you put insulation from wall to wall. This had a special type material they put in there—I didn't even know what that was. There's a lot of cables in there, and other various things. I never saw anything like: all of a sudden a section of the ship became clear. We didn't do any of that.

Ques.: The *Manzano* base in New Mexico, there's been a longstanding rumor of alien involvement there, and occupation at deeper levels. Is that true, and have you been in that facility?

B.U.: I've been in that facility. I know they say White Sands, but it's not on White Sands. There is a tunnel from White Sands *to* that facility.

"I had a funny thing happen to me a few months back, 3 or 4. I had a call from this professor at Stanford [Univ.]. He knew a lot of info. about me, where he got it I don't know. He wanted to talk, and said: 'I want to talk to you about a problem we're having, and I understand it's OK to talk to you any time you're around...' I said: 'well, just be sure and let me know. [But] I need to know who you are.'"

"He sprung this on me, just like anybody else. They call me up, they say 'Hey, I'm here, and I'll meet you at a certain place, but I only got 3 hours in town, can you come?' Well, I was interested in the guy because he sounded Hungarian, right? His problem was, apparently, at Stanford they got a test cell; one that's actually built in miniature configuration. And they had problems with turning it; making it go left and right and calibrating it. So, he knew that much—that I worked on avionics. He wanted to know how we got it to turn. I said: 'Why don't you talk to this guy in San Diego, he also worked on it. Besides he's instrumentation and control, I'm just in mechanical.' He said 'No, I was told to talk to you.' So, see, there's a method in the calibration, there's steps you gotta follow; particularly where a plane is at the end of an airway—what do you do? You calibrate it yourself. Well, apparently this Hungarian went away happy. He was a Hungarian, I knew that. I asked him that!

Ques.: How many craft do they have altogether?

B.U.: I know there are 3 simulators, I know that. But that's in 1989. I'd say, of craft in different configurations, I'd probably say there's a good fleet of them, maybe 50 to 100.

Ques.: With an actual operating craft—do they have a lot of similar characteristics, as far as effecting other power systems? Such as, you're driving a car and at close range the craft *that we build* could interrupt the power of a car, or perhaps a house? Do you know that?

B.U.: Sure, I think it would; I think they do. We did have—when this was cranked up—we did have instrumentation. I can't go into details about it, but we did have instrumentation to identify just what the effects of the field was. From the center to, say, 30–40 yards away.

Ques.: Would it affect a human being too? Could it be harmful?

B.U.: It could, yeah

Ques.: The actual ships being flown, what are they used for?

B.U.: The ships, they can't carry very much. Basically, it's just surveillance. [They're] still training; still trying to understand it as far as humans are concerned.

"But let me say this: we took the simulator, and it works just like a craft. Everything had to be in perspective, exactly the same [as in the craft]—weights and everything else about it. The people are part of the whole thing [as an integrated system].

"If they build one, they're gonna have to make sure that the power plants for them—they're gonna have to build it as big as a football field: with the engine and system for operating it in the center; and people on the outside, shielded from all the bad stuff.

Ques.: How do they charge the craft? Are they self-contained, so they don't need [an outside power source], or do you do something from the outside to give them an initial juice, or jump-start it, or what?

B.U.: The capacitors, we charge the capacitors. On the simulator, we had the 6 capacitors which were one million volts each. Now, we could run for 30 minutes on that 6 million volts. Now, you're gonna ask me about what they did in the actual craft: As far as the generator that was gonna generate the voltage to operate that ship? I have no idea.

"I know this much: I didn't have anything to do with it [i.e. 'need to know']. But they did have a unit that sat below it, and just above it; and this shaft that ran up to the top of it that was a waveguide type thing; it wasn't just a waveguide, it did something else too. I put my hand on that. That was not part of my function: you only did your own job, but I did put my hand on it [the waveguide shaft], and it's *very cold.* It's a very cold pipe, so it's a cold fusion type [system]. If I was gonna tell you anything, you know, my feeling and my analysis of it? It's a cold fusion reactor. And they're using material that can not decay as fast as regular nuclear stuff. Even the *element 115* that Bob Lazar talks about is dangerous. They should respect that if you have one blow by chance, you could have probably killed a lot of people—which would be a bad idea.

Ques.: What is it, a terrestrial element?

B.U.: It's a terrestrial element, yeah.

Ques.: One on the periodic chart we could recognize?

B.U.: Well, you know, the one that the bureaucrats are gonna give you, that's not the actual chart! You have to go into that [systems briefing] book I told you about—*FOGIT*—then you'll see how far it goes. And I'll tell you what: I saw that chart, you... what was the latest one you saw? Tell me the year [of your reference]?

Ques.: [Up to element] #104, 1962–'63.

B.U.: In 1962 I was looking at 121 [elements]. I didn't know all about it—I'm not a chemist. I'm in mechanical-design engineering. I don't know anything about that part of it, but it's further [enumerated] than you think! If you ever get a chance to look at *FOGIT, and you're able to tell about it,* you'll be...How long [have] we had 104? A hundred years?

Conservatives Get It Right, and Liberals Don't Get It!

Ques.: The 6 million volts that powered the simulator: Did they identify any side effects from being in that simulator with that kind of voltage; electromagnetic fields, or...?

B.U.: I'm 72 years old! I went through the operation 30 minutes at a time. So, I probably spent 100 hours; a hundred times 30, by the time we were all done...I'm still here!

"Now! Who's it going to effect? It's probably gonna effect somebody with bad genes; they're not physically, you know, up to snuff. It might make somebody blind or something, or give them cancer like smoking and that kinda thing—but who knows?

Ques.: Are those aliens contagious?

B.U.: No, no, they're not contagious. They smell a little bit, but like we all do.

Ques.: What about the possibility of psycho-kinetics being involved in all of this—mind over matter?

B.U.: Kinetics? I believe there's a lot of things you can do with your mind, you know, that you've gotta train it. It's just like I said earlier: In order to train pilots in this thing, you gotta take them and re-orient their brain! So, that's training of the brain.

"...There's a lot of people that are denying it, and resigned [to the old mindset]. They're not even—they don't even want to cooperate! I mean, how many people in this room—you know, if you get a group like that—how many are gonna be able to [link] minds; make a mind connection, so to speak? That kinda thing. That's gonna be difficult. If you start off with a few, kind of work up and bring them in, you might have a chance.

"The ones you'll have the problem training are the *liberals!* So...I'm not a liberal. There's nothing wrong with the earth; the ozone field or anything else. That's just more bureaucracy, to scare everybody out of their wits. We gotta do this, or methane from cows...they say, 'hey, a big asteroid destroyed the earth', you know, destroyed the dinosaurs. They're saying we're destroying all the ozone with all the Freon, from putting it up there—well I think about all those [mega] dinosaurs that were poopin' on the earth, and gassin' off and all that stuff [for 165 million years]! Worse than methane, *worse* than volcanoes and stuff.

Afterword: The Candor of a Cover-Up?

As a public offering, the Uhouse case is uniquely credible. It's the matter-of-fact simplicity of delivery that brings accessibility to his disclosure. And yet, the panorama of circumstance volunteered by Uhouse, presents an untoward paradox to advocates of ufology. His reasoned candor affirms the pithy maxim: "Truth is stranger than fiction." Although, with Uhouse, the *high-strangeness* is not so strange after all. His paradox is how the comprehensive bearing of his account is actually so plausibly down-to-earth. At least, the respective "need to know," which Uhouse was obliged to accept, both informs us and certainly belies the fantastic imaginings of what coexistence with space aliens may actually mean. In its most practical translation, the credibility of Uhouse becomes exceptional precisely because it simply makes sense. The enterprise of his alleged career seems easy to understand, and is internally consistent.

Thus, the internally consistent details, according to Uhouse, are congruent with related correlations from a history of other sources. The question is, whether the paradox thickens due to his candor? Why was Uhouse allowed to reveal his story? Those very few ufologists who studied the Uhouse disclosure, know he was very candid in claiming his supervisor—and/or personal security officer of over 20 years—allowed him to disclose *some* of what he knew. This is why the Uhouse case is totally unlike the Bob Lazar case, about the same basic proposition: human engineering of flying discs at the S–4 site. So, Uhouse would appear not to be so much a controlled "leak" as an unofficial discloser. This is curious, since his candor contradicts the perception there is an official "cover-up." For many years it was perceived an official policy existed, accurately defined as a "cover-up"—now evidence actually contradicts this, as how the system *actually* works.

The Uhouse case would, in this light, give good cause to rethink the ongoing vetting of the *Majestic Twelve* documents. Regarding unverifiable *motives*—are these docs only available due to the good conscience of some well meaning agent? Or, are they a controlled *leak*? Or, are they an unofficial disclosure? The Uhouse case suggests some sort of managed disclosure model—not the inadequate "cover-up" model. Quoting my collaborator, Melinda Leslie: "It's a matter of semantics, but the semantics are drawing the line."

It actually seems to be, that if the medium of the Uhouse message is his candor, then the medium (candor) is his message, and not the content of his experience. This is likely, because according to the "cover-up" thesis, the Uhouse set of classified specifics could never be allowed disclosure! The very unique quality of Uhouse is that his account converges those factors of the UFO enigma to which the public would most easily relate to: human affaires pursuing a utilitarian application of allegedly ET-derived concepts.

In other words: we have nearly 60 years of data collection. Our obligation now may be the need for a different model to explain all the facts about how the public is informed. Any circumstance involving factors totally unpredictable and uncontrollable—i.e. ETs, "leaks," etc.—should properly be described as a managed informant campaign, managed for damage control. The fact is that Uhouse *did* unofficially disclose for the years 1994–98. Certain details he said he could discuss, certain details he could not. So, the candor of the Uhouse briefings do not conform to a "cover-up" mentality. How, then, do we apprehend this?

Throughout this ET data-gathering era, there has been an impediment to fully comprehending the total body of evidence: It was an additional demand placed upon the civilian researchers *by* the evidence; the military involvement and its prerogatives of intelligence. This demand can never be thoroughly satisfied, since civilians generally have no means for gaining access to the role played by Mil-Intel at any point in time.

There has been an effort, however, to better illustrate these implications—how the role of Intelligence may be influencing the human response to the ET phenomena. One good example is a book titled *UFOs and the National Security State,* by Richard Dolan.

This author defines the National Security State as an entirely independent apparatus which acts by its own rules, beyond even Presidential control; i.e. "a state within a state." This "state" is actually a construct for decision-making, not bound by the same laws the rest of us are bound by.

Some political researchers perceive this extra-governmental "state" as the *Shadow Government:* you might say it's a consortium of corporate forces, who wield the power of corporate wealth, in pursuit of policies rooted in doctrines of undemocratic ideology.

If it seems we are drifting away from Bill Uhouse, we are not. Of the many revealing assertions Uhouse made, the one which may pose the greatest disconcerting insight to ufologists is the following: his construct for repeatedly distinguishing the organization who employed him for 38 years: "I want everybody to understand that It's Not The U.S. Government!...they do have, essentially, a government that's in a box, that's separate from the U.S. Government...I'm talkin' about a different government: a *Satellite Government!* It's not the same as our Shadow Government *or* the U.S. Government; it's a government in its own right!" This proposition, if accurate, means the structure of the system ufologists perceive they are dealing with has been a diversion. And, in midstream, the game changed venues without informing anyone. That is, until Uhouse temporarily surfaced! Not ironically, just before Uhouse stopped disclosing, Col. Corso released his book and explained the same scenario. Corso said, *MJ12* "would form nothing less than a government within the government...". Coincidence?

Richard Dolan has been well received for introducing the protocols of our National Security State into UFO studies. I mention this, in light of the model that the National Security State apparatus sets forth, as an expedient for *extra-governmental* affairs. This "security state" principle of organization, is what Uhouse infers—by association—when he emphasizes the clear distinction between our Constitutional Government, the *Shadow Government* of doctrinaire ideologues, and his own employer—the manager of the ET/UFO/reverse-engineering nexus: "I call it the *Satellite Government!*" In other words, the *Satellite Government* is an embedded subordinate: clearly powerful, independent, and not to be confused with other familiar machinations of command & control. All this makes sense! It is consistent with the reverse-engineering scheme defined by Col. Philip J. Corso. Our government doesn't do R&D work, it farms R&D out to the M.I.C.

Given all of the above, the "cover-up" may not be so much about refusing to officially confirm the ET-presence. The "cover-up," as a part of a system, serves to avoid publicizing complicity in saucer manufacture, triangular craft manufacture, etc.

The convergence of data according to both Corso and Uhouse provide perspective—human scale—to be applied to ufology. A human scale perspective, at this point in time, is the bottom line. In our current paradigm, the bottom line is the same, regardless of the issue. Corso and Uhouse both emphasize this: the ET-technology/info is a fortuitous resource base for sci-tech innovation, and product development. This nuance of detail is exactly what Uhouse stated was the payoff for the M.I.C. companies who have invested in the ET trade pool. This is why this section is called *Ace In The Hole.* Forensic ufologist/illustrator, Bill McDonald, summed up the "cover-up"/disclosure data with the same premise: "If you follow one of the number one rules in the detective's world—which is 'follow the money'—economic prosperity for the persons involved in the issue of the secrecy, is directly linked, directly linked to their ability to keep their mouth shut; and it's *directly linked* to the entire 'cover-up.'" Obviously, a plethora of paradigmatic issues exist in redefining a *reality* wherein ETs would take historic precedence, even on earth. In the meantime, corporate dominance and technological advancement set the agenda!

As an abductee once told me: "The ET phenomena is a human phenomena, before it's an ET one." On his July 20, 2002 radio broadcast of *Dreamland,* Whitley Strieber rendered the status of UFO research: "This *is* important—we *have* enough proof! We *know* something is going on." Strieber is correct. So, if we have enough proof, then the approach to articulating the issue should reflect an accurate perception of how—and by whom—the "proof" is being managed. The issue now is not gathering more circumstantial evidence. The issue, also, is not that concerned citizens should petition governing bodies to disclose their complicity in something that—the evidence indicates—has long ago been safely privatized for developmental dominance.

The issue should be, that the UFO community consciously collectivize and correlate its data base into a coherent basis for institutionalizing a different reality; a paradigm inclusive of a historic ET-presence on earth. Evidence indicates, the old but prevailing paradigm dictates there is only one disclosure style presently feasible: the "cover-up" is the disclosure, and disclosure is the "cover-up."

** Special thanks to Shawn Atlanti for his assistance in placing the testimony of Bill Uhouse on the record!

Bill Uhouse publicly presented his accounts in: May and Sept. 1996, May 1997, and May 1998 in Rachel, Nevada; presented here in abridged form. These lectures/dialogues by Bill Uhouse were recorded originally by Shawn Atlanti, and transcribed here by Randy Koppang.

Graphic layout and audio recording enhancement by Free Spirit Productions.

Appendix A

Immediately prior to publishing this Bill Uhouse testimony, an invaluable addendum was volunteered. This resulted from a synchronistic opportunity facilitated by respected investigator Bill Hamilton. Thanks to the advent of internet communications, Hamilton established a rapport with the son of Bill Uhouse. Hamilton then met with him, and this author spoke with the son, Will Uhouse.

Will is known by both his given name, Will Uhouse, and his cyberspace pen name Will Sheephogan. The following details were contributed by Will for bringing greater coherency to the account of his father's career.

Will's claim of vivid memory goes back to when he was three years old. His father would take unexpected "business" trips to debriefing sites such as Colorado Springs, Colorado. When Will was this age, he said the family resided in Cape Canaveral, from 1955 through 1963. However, a routine obligation of his father's position required him to periodically reside at sub-rosa facilities for weeks at a time.

Throughout his career, Bill Uhouse actually served in a multi-purpose capacity: He worked for high profile M.I.C. companies such as Aerojet General, Westinghouse and others. Yet, simultaneously, Uhouse performed duties for an "underlying company", as Will described it. When the need arose, Uhouse would be pulled out of the overt jobs and into the project sites for extended periods.

Will offered an insight regarding the inferred dual, or multi-proprietary structure of this arrangement. Will perceived that the covert level project was pursued by an internal corporation, utilized by all the known contractors. This sub-company was created by the mainline collective for supplying varying degrees of expertise on the project. This, of course, is a privatized, highly compartmentalized "company". It is managed by what Bill Uhouse called "the satellite government"; a structure that no one member of personnel can put their finger on.

In answer to the question as to when his father first spoke about any of this—to a group of outsiders—Will's experience settled on 1967 or '68. He remembered a library type gathering in California: The audience was invited (not open door public), with brothers of Bill Uhouse present. Will claimed to have a memory of attending this meeting, but as a kid going in and out of a room of 40–50 people. Currently, Will feels new developments have drawn his father back into the program. This would likely be in an advisory mode, thus obligating Uhouse to become unavailable (as he is now).

To the degree that the R&D of a human replicated disc-craft is pursuing a better, slower, stealth aircraft, the corporate involvement precludes official agency disclosure. However, Will's experience concludes that this arrangement ultimately allows for some information becoming available. The Bill Uhouse disclosures may be an exemplar demonstration of this.

The private sector provides for different relationships with people who work on this. The private sector must work with contracts and timelines, about which the program is discussed. In the military, classified info does not include contracts and timelines (as do industrial applications). Personnel relationships pursuing the corporate learning-curve require information flow. So, information has been known to filter down. When personnel retire from their fields, and their contract of privacy is over, they may feel comfortable in divulging info not the most crucial to national security. And, as in the Bill Uhouse case, they may be secure in doing so. Uhouse confirms this form of disclosure—simply by his presence alone—as a source.

Publicity about the stealth fighter or the B 2 bomber became openly known through a complex of unofficial and official sources. And the pattern may apply to the disc/triangular (stealth) aeroforms now.

Will offered an insight into the triangular or delta-levitation craft reported in Nevada skies: He observed a facet that reflected the stealth design of the delta-craft after two daytime sightings. Clearly, they can fly too slow to remain aloft, and they make virtually no sound. But the delta-craft Will observed in Stewart Valley, Nevada was not "black" as commonly described; it was actually a mirror finish. At a certain angle of flight, the wing disappeared in camouflage as the mirrored reflection merged with the mountainous terrain aligned with the observation. Then, the wing (outline) reappeared as the maneuver changed the angle of reflectivity and the image of resolution. (Does this mirror quality explain more accurately the common description of these craft as being black? In low light or at night, they would be black, but black is not a very stealthy daylight color. Plus, the UFO model of levitation-craft suggests the propulsion system may dictate that no painted color is desirable.)

Will felt these sightings were an important reinforcement for what his upbringing brought to his attention. Yet, he imparted to this author an authentic sense of experience that dispelled these issues as being extraordinary. For Will, all this was something he grew up with since being a young child. Will perceived the otherwise astonishing as normal "everyday stuff". He illustrated this with one memorable incident: His father arrived home from a debriefing and shook his head, exclaiming that 'they're real'; that we have one of their craft, and he was going to be working with it. This moment seemed to set the mood for Will's upbringing. Revealing this involvement to the family was dutifully practical; Will felt it volunteered an explanation for why Bill Uhouse repeatedly "disappeared" for a few weeks at a time.

In this way, Will learned that the program was located at the Nevada Test Site, at Los Alamos, at Colorado Springs (perhaps the NORAD complex there?), a whole variety of locations. In other words, the collective body of data—indicating the levitation-craft project exists—would predict a complexly faceted program, requiring many specialty sites. The non-disclosure or "cover up" question is most pragmatically solved by this possibility. The many sightings of anomalous delta-craft—like that of Will above—seem to be tests; perhaps born out of the reverse-engineering data. This, in turn, could lead to a better, safer mode of space travel.

The "cover up" would protect the perfection process. The question is: how far along is the program? Will does not think it is as far along as might be presumed. The issue why this may be the case has to do with, not so much the hardware, but the psychic-wetware (i.e. *the mental-interface* with navigation). The collective ufological data has always pointed to this, as Melinda Leslie explains.

If there are ETs, we are not them. So, when will humans be competent enough to imitate ET modes of conveyance? As Will observed, by imitating ETs we build understanding. But, if we possess an alien model for such a quantum leap, we might also need to speak with they who made it. What is their *extrasense* allowing them to be merged with their craft? (It appears they are an integral part of the craft, without some mechanical interface that humans require).

Answering this question is crucial to various areas of investigation in ufology today. It is a connecting link for bringing coherent logic to the themes of continuity emphasized in this book. If there is any validity to perceiving a quantum leap in human understanding of ETs, the issue of mentally-interfacing with technology surely symbolizes such futurism.

In the testimony of Bill Uhouse above, he says: "If you think that we're at the end of science and technology—we're not! We're just beginning!" Receiving official confirmation of ETs, however, may depend on this issue of technological direction. How fast can the public tolerate a learning-curve of technological progress inspired by ETs? Why would an Official Disclosure already have been made of something science insists is impossible? Such an ambition as replicating an ET-technology—requiring the user to psychically-integrate with the operating system itself—contradicts all our assumptions about humans determining the direction, or destiny, of our own future.

What seems to be occuring is the pursuit of a technology whose optimal viability challenges the user to impart a directive will, by mentally-interfacing with it: Is this too great a challenge for both technological definitions and postmodern culture alike? As with the concept of reproductively cloning humans, the mere possibility may alienate fundamentalist values. Mental-interface technology redefines our co-dependency with Artificial Intelligence.

An ET derived technology will dissolve the boundaries of human traditions and identity! A study of media (technologies) will tell you this. Officially Disclosing ETs removes the perceived doubt regarding the existence of the reverse-engineering that Col. Philip J. Corso defined.

Yet, such issues are now becoming more plausible as the product of ufology today. These studies also illustrate our own paradox of awareness. As an astute observer of creative progress has recently said of our time: "It is a time when everything seems possible but nothing is." An observation of this nature is indeed a sign of our time. The paradox of contradictory opportunities is the tension of a shifting social paradigm; transcending our pastimes of past times, and shapeshifting culture so as to symbolize a broader sense of identity.

Addendum of Anticipation

On November 10, 2005, I was told that the above Bill Uhouse testimony was timely, because the long awaited official "disclosure will be made next year"! An official confirmation of ETs would afford validation to the unofficial disclosure by Uhouse. This anticipation was offered to me with doubtless confidence: The source of this forecast is a well respected ufologist investigating "insider" leaks and underground M.I.C. facilities. Identity of the source(s) is withheld, so as to ensure bias-free objectivity about such a forecast.

Fortuitously, this anticipation of such a breakthrough presents us an opportunity to do some science. Will the planet Earth receive an Official Disclosure of the ET-presence in the near future? Notwithstanding all the qualifications above—for whether to expect a near term ET disclosure—one determining factor is clear: The multifarious motives defining the inescapable political (or exopolitical) impositions of our day do exist. So, the reality created for believing (wishing) that such a disclosure will occur shall also demand such a revelation to serve a purpose; an agenda; a set of contingencies. The science of human (political) behavior is predictive of this fact, as a context for any disclosure.

Appendix B

Drawing upon our ultimate set of values for redefining an ET cover-up, the onus is on us. Only the UFO community can bring an unbiased disclosure of the ET-presence directly to society. Yes, ufology may become more scientific. But institutionalized science will not aid in *disclosure*. This is because *"the model for what science should be"* will not—cannot—respond to the *nature* of ufological phenomena. Today, publications on the issue should, therefore, be as comprehensive as possible.

With this in mind, *Appendix B* magnifies our hindsight regarding the historical record. Two developments are historically central to our redefinition: human/UFO hostilities, and ET-involvement with humans.

The earliest attempt to correlate these two possibilities, I have located in popular media, was of May 1957; not in the terms used today, of course.

Pioneer publisher Ray Palmer was to present these possibilities, prior to the very first national magazine on the UFO enigma, in June 1957: *Flying Saucers From Other Worlds*. Palmer originated other magazines such as *Fate* and *Search*. In his May 1957 issue of *Search*, an article title asked, *Are There Cosmic Kidnappers?*, by Alex Saunders. This article catalogues a series of historical anomalies. Not ironically, they anticipate the first mass media reportage of a landmark "abduction" case, in 1966. This was the Betty and Barney Hill case; the case appeared in *Look* magazine, on the *Mike Douglas* TV-talk show, and in John Fuller's book, *The Interrupted Journey*.

Why such media exposure was a landmark is explained by Jim and Coral Lorenzen (APRO), in their 1967 book *Flying Saucer Occupants*: in 1966 there had been "a strong resistance to the idea of UFOs being occupied by men ever since... 1947" (p. 84). The Hill case was evidence for aliens.

Thus, nine years prior to the Hill case, Alex Saunders accurately deduced such incidents as reportedly occurring. In *Search* magazine Saunders asked, *Are There Cosmic Kidnappers*?

This same article voiced another major concern of today. It is a scenario proposed by Linda Moulton Howe (via Andrew Kissner), regarding possible UFO retaliation against human offensive acts towards UFOs. In the following, Saunders anticipates such issues, currently a primary trend in the UFO studies of 2006-7: i.e., explaining evidence for numerous UFO crash-retrievals. Saunders reports: "Heavily armed military planes with up-to-date radio facilities have ceased to exist after leaving their base! Whatever befell them did so with blinding speed.

"Sometimes the wreckage of a plane is found, but *only* the wreckage. Bodies of the crew or passengers are strangely missing, whether the "accident" occurred over desolate waters or densely populated terrain. Mysteriously, planes explode in the air, crash on mountain sides."

These data correlate to both *abductions*, plus the hostility/retaliation thesis of crashed UFO scenarios—but at a time when researchers resisted contactee evidence that ETs occupied flying discs.

Today, we have a nearly identical resistance among researchers. My objective here carries forth the forensic model stated by Jim and Coral Lorenzen in their book *Flying Saucer Occupants*. In their forward they state their intentions are "to deal with aspects of the UFO problem which have been for the most part overlooked or sidestepped." The parallel in 2006-7, is a perceptual resistance to assigning explanatory meaning to substantial evidence for human intervention/monitoring of ET experiencers. As detailed previously, this logical human resource of evidence holds the only explanatory power for clarifying the coverup-disclosure; and the role ET-witness reabduction plays in official *non*-disclosure.

In anticipation of "abductions", Alex Saunders concluded his 1957 *Search* article, saying: "And so it goes. Year after year. Possible levitations of the living and non-living...

"*Like cosmic kidnappings*?

"If an alien race from an advanced world is responsible for missing Earth people, what, the question is again asked, is their purpose? ... Perhaps the answer cannot be presently guessed." Yet, today we can.

Our contention—at this stage of apprehension, or research—is that if a few hundred alleged abductees provide a consistent pattern of human involved evidence, it should be rigorously pursued. I.e., Mil-Intel-appearing humans reportedly are forcibly interrogating witnesses about their contact biography. Such human monitors are the most logical sources for answering the above questions from 1957, and today! But presently, abduction research appears to resist this forensic logic.

Affirming this stance is not new. And yet, in this very area, of investigating an Intel-involvement with abductees, it proved formative for redefining an ET cover-up.

At *The UFO Experience* Conf. in New England, October 1995, Leah Haley and Marc Davenport addressed these issues. Davenport assigned unforeseen meaning to a test of Leah Haley's experience: she, and two others, successfully transited 13 miles of a security restricted sector in Eglin AFB, Florida. Haley wanted to photo-confirm having possibly witnessed a crashed saucer there. The test was to repeat her intrusion into Eglin.

In the test they were intercepted by security. So why was Haley's first entry clearly allowed and observed by security? Davenport reports their deduction: "The point is that Leah was allowed to walk here [into Eglin] the first time. It seemed to be sanctioned somehow. The second time she wasn't. So what does that mean? To us, it means that there is some kind of acclimation program in effect here [given

Haley's bio of Intel monitoring]. And that some people, at least, within the government want this information out. Even though some people still want to keep it secret...

"It appears now [1995] that there is an acclimation program in progress. Maybe it's been in progress a long time, maybe since the Roswell Crash... But it appears that there are at least two factions [involved]...

"In Nov.-Dec. 1994, Leah and I were privileged to be at the *Mesquite Conference* in Nevada... and along with 18 other researchers, we were asked to participate in a roundtable discussion with a civilian representative of the CIA... [This author confirmed this meeting as actually occurring, in detailed conversation with *Mesquite Conference* producer Bob Brown.]

"This representative told us — he said, 'Don't expect the CIA, or any other agency to make an announcement about this.' He said, 'You're going to have to do that work; announcement to the public.'

"The same man later told me, 'If you have the right questions you may get some answers, now,'" (1994). Davenport then deduced that "the right questions" are about the government projects that were conceived specifically to recover crashed flying saucers, (i.e., Operation Blue Fly and Operation Moondust).

Perhaps it is not ironic that 12 years later, the subject most compelling in ufology of 2006-7 is a topic now having its very own conference: the crashed saucer/retrieval enigma!? And as we illustrated earlier—release, or disclosure, of the crashed disc-hostility scenario could be a factionalized setup, a psy-op, to ultimately devolve ETs into the ultimate post-Cold War threat...

In Sept. 1992, *Fate* magazine interviewed abduction authority Bud Hopkins.

Fate asked the question: "Have any of the abductees with whom you've worked, or any other abductees whose accounts you've heard from your fellow investigators, given any credence to the tales told by John Lear, Bill Hamilton, and Bob Lazar about underground alien bases, treaties between aliens and the U.S. Government, and so forth?"

In terms of verifiable facts, this question was sensational, and phrased imprecisely. Bud Hopkins replied: "Not a single one. I've been doing abduction research for 17 years and I have never heard any of this from any of the abductees I've worked with. If it's going on, I would have heard about it." Shortly thereafter in this interview, Hopkins makes a qualified statement in answer to, "how long abductions have been going on"? He said, "...I can't be sure of that because what you hear about isn't necessarily what is exactly out there." This caveat would have wisely been applied to Hopkins' lack of knowledge about underground facilities—as opposed to alien bases. Alas, Hopkins' opinion was not an informed opinion. If no witnesses (abductees) are reporting underground experiences to Hopkins, then it was not a datum pertinent to his inquiry. Thus—*what we do not* "hear about isn't necessarily what *is exactly* out there" either. In other words, the converse of Hopkins' caveat also, fairly, applies.

However, in the 14 years since Hopkins' interview, unprecedented sources of evidence have surfaced indicating the predictive logic of why M.I.C. underground facilities play a crucial role in the UFO-info-management agenda. Not alien bases. Rather, military-industrial contractor underground bases. Those used for pursuing crashed saucer security.

Today, a significant cohort of abductees claim also to be monitored and interrogated by agents, about their perceived ET-contact experience. Just like the circumstance whereas researchers resisted the idea there existed UFO occupants—today, opinion resists the evidence that certain abductees are being monitored by humans, about their contact with UFO occupants. And these witness accounts should not be misperceived as humans impersonating ETs, and staging horrific abduction trauma. The compiled pattern of evidence leads, instead, to the crashed saucer LCD throughout all ufology.

Ethically, the "re-ab" case histories have system-wide implications as a human rights-type path for ufology. Given this, do I expect a major shift will occur in the field of ufology, towards this human-human set of correlations? No. I certainly do not. All I am actually saying is that—in terms of forensic logic—this known set of evidence is where the human response to the ET presence can only logically lead. And coincidentally validating this logic are American current events of 2006-7. These are instructively similar in terms of human rights.

A public outcry is growing now over: non-warrant phone-tapping by the National Security Agency; indefinitely imprisoning suspicious immigrants (aliens) without charges; tortuous interrogations by Mil-Intel; using totally unproven national security agendas for starting the Iraq War, etc. In their own way all these current constitutional affronts apply to the principle of procedures used in monitoring or re-abducting abductees against their will. If citizens cannot influence such current events in the interest of good government—citizens will not penetrate the National Security State priorities of how ufology has been managed (ala Col. Corso's explanation).

Ufology is a social education process, not a government oversight committee/crime investigation. Ufology is managed in terms of how its best evidence has largely come to it: unofficially, from military-industrial ("insider") sources.

In the most practical terms the greatest social value ufology can offer is to collaboratively assume its role as a conscious medium of change, as a harbinger of paradigmatic change. The evidence is that this is ufology's actual message: ufology as a vector or medium of change is its message. For however long it takes, until society at large is totally *aware* of ET, pro or con. And this awareness is the Keystone for co-creating a different global cosmology. "And *now* for something completely different!" The message is not the message, "the medium is the message"...

Index

Printed in the United States
105371LV00003B/318/A

9 781585 091102